Theology in Science

Kenneth Francis

En Route Books and Media, LLC
Saint Louis, MO

En Route Books and Media, LLC

5705 Rhodes Avenue

St. Louis, MO 63109

Contact us at

contactus@enroutebooksandmedia.com

Cover Credit: Sebastian Mahfood

ISBN-13: 979-8-88870-501-8

Library of Congress Control Number:

Available online at https://catalog.loc.gov

Table of Contents

Introduction

In the secular, hijacked West, the so-called 'Intelligentsia' (should be called 'Dumb&Dencia') and atheist technology tsars will often say science and theology/philosophy are incompatible. In his co-written book with Leonard Mlodinow *The Grand Design*, Stephen Hawking makes the mother-of-all self-refuting statements by claiming 'philosophy is dead', the latter being a philosophical statement. But it gets worse: He goes on to say the universe created itself (I am not making this up). According to Hawking, "Because there is a law such as gravity, the universe can and will create itself from nothing." I raised this error at a science conference in Belfast shortly after the book's publication.

Chapter 1

God and Quantum Theory

In understanding the universe, it seems that quantum theory, according to most physicists, is the final mysterious frontier of cosmic science. If so, I believe that this boundary can only be understood fully by a Mind possessing omniscience: God. The serpent in the Garden of Eden successfully conned Adam and Eve into believing they could achieve omniscience, and we all know what happened after that mother-of-all conceited errors.

I often wonder did the serpent also tempt the 'Adam and Eve' scientists in Switzerland's 'Garden of CERN', with their Large Hadron Collider in search of the so-called 'God particle', with the company's abstract '666' logo, and the Shiva statue god of creation and destruction outside their creepy headquarters. Even Mary Shelley's novel *Frankenstein* was written by the shores of Lake Geneva, next door to where CERN was later built, with its potential for creating another Frankenstein's monster.

Omniscience entails knowledge of everything, including how the universe exists on an infinitesimal micro level. The late Stephen Hawking, who was an atheist, once said: "If we find the answer to that, it would be the ultimate triumph of human reason—for then we would know the mind of God [figuratively speaking]."

So, what is it that is so complex about quantum theory, and how is it related to God? Quantum physics is the study of matter and energy at the most subatomic levels. However, scientific aims to uncover and fully understand the properties and behaviours of the very building blocks of nature seem to be halted by a metaphorical brick wall, like the Hand of God saying: "Back off! Remember the Tower of Babel and the conceit of Man trying to reach into Heaven? Like Babel, I've no desire to fully destroy the Tower of Quantum Theory, but I will frustrate the communications of scientists in their quest for omniscience. Only one who possesses omniscience can fully understand this mysterious field of knowledge. But I will allow you a hint of it while studying my Resurrection and the Shroud of Turin, as well as some other aspects of science." (Apologies to God for speculating on what He might say!) And, theistically, isn't God with His omniscience and omnipresence combined the only one capable of solving the quantum paradox of Schrödinger's Cat, by knowing if it is dead or alive?

Cats aside, the few things that scientists theorise on this weird field of cosmology are undeniably spooky. In this world of small particles, classical physics is no longer valid and seems to break down and act strangely. This is unlike our macro world, which describes the movement of big particles, where everything is deterministic.

But in quantum theory, according to physicists, there are 10 interpretations, thus the big problem lies in interpreting the equations of the mechanics: Which one of the all-valid, all mathematically consistent 10 interpretations is the right one? On the quantum level, we can only determine probabilities for the movement of the smallest micro particles. There is also the hypothesis among some scientists that the brain, not the Mind, relies on quantum physics.

Writing in the online science magazine *Mind Matters News*, author-journalist Denyse O'Leary says: "[Such a] hypothesis is that the brain relies on quantum physics, not classical physics, to power thinking processes. Quantum processes are helpful to know about when we hear a gimcrack new theory that dismisses or explains away consciousness. We know it can't be that simple."

The Irish philosopher George Berkeley's Idealism touches on the Mind and consciousness. In *Of The Principles Of Human Understanding*, he wrote: "It is indeed an opinion strangely prevailing among men that houses, mountains, rivers and in a word, all sensible objects, have an existence, natural or real, distinct from their being perceived by the understanding … For what are the aforementioned objects but the things we perceive by sense? And what do we perceive besides our own ideas or sensations? And is it plainly repugnant that anyone of these, or any combination of them, should exist unperceived."

Bible teacher Chuck Missler[1] wrote about a Hungarian particle physicist called Dame Isabel Piczek,

---

[1] "A Quantum Hologram of Christ's Resurrection? An Easter Surprise," (April 1, 2009), *Koinonia House.* Online at https://www.khouse.org/personal_update/articles/2009/quantum-hologram-christs-resurrection

who researched the Resurrection of Jesus Christ. Missler said she uncovered hard, scientific evidence that Jesus Christ did, in fact, rise from the dead. The object, which was wrapped around the body of Jesus, of her study is a simple piece of ancient fabric known as the Shroud of Turin, the most-studied artifact in history.

Dame Piczek explains the complicated physics behind the image on the Shroud: "As quantum time collapses to absolute zero [time stopped moving] in the tomb of Christ, the two event horizons [one stopping events from above and the other stopping the events from below at the moment of the zero time collapse] going through the body get infinitely close to each other and eliminate each other [causing the image to print itself on the two sides of the Shroud]."

She and a team of hardboiled scientists (many who allegedly converted to Christianity) also discovered that the body of Jesus defied gravity by elevating, while the Shroud became taut (like a sideways letter 'U' shape), as He levitated then floated in the middle.

News of this went mainstream in December 2011. The UK's *Daily Mail* wrote in its headline:

Turin Shroud 'was created by flash of supernatural light': It couldn't be a medieval forgery, say scientists.

The article continues: "Italian researchers have found evidence that casts doubt on claims that the relic—said to be the burial cloth of Jesus—is a fake and they suggest that it could, after all, be authentic. Scientists from Italy's National Agency for New Technologies, Energy and Sustainable Economic Development spent years trying to replicate the shroud's markings. They have concluded only something akin to ultraviolet lasers—far beyond the capability of medieval forgers—could have created them."

As someone who is self-educated in astronomy and neuroscience, I cannot verify the above findings, but I can speculate on the core aspects of these findings based on reliable sources by both theistic and secular distinguished scientists whom I trust, as they are not ideologically driven by the corrupt field of Scientism.[2]

---

[2] Cf. Henry Miller and Stanley Young (Jan. 29, 2024), "Part 1 – Viewpoint: Why is Trust in Scientific Research at an All-Time Low?" And a Pew study by Brian Kennedy and Alec Tyson (Nov. 14, 2023), "Americans Trust in Scientists, Positive Views of Science, Continue to Decline."

To repeat: In the strange field of quantum mechanics, if it does exist, only God understands how ultimate reality works, leaving us to settle and be content with some clues of this alleged mysterious micro-universe. Let this be a warning to those who seek omniscience. Man is born unto trouble as the sparks fly upward. (Job 5:7)

Chapter 2

When AI is No Laughing Matter

There has been some talk recently on the AI (Artificial Intelligence) grapevine about the wonders of robots analysing data and performing literary tasks far superior to humans. Although I don't trust mainstream-media propaganda anymore, I'll give this report extract below the benefit of doubt: According to CNN, an AI chatbot is dominating social media with its "frighteningly good essays."

The report said that OpenAI opened access on November 30, 2022, to ChatGPT, an AI-powered chatbot that interacts with users in an "eerily convincing" and conversational way. Its ability to provide lengthy, thoughtful and thorough responses to questions and prompts—even if inaccurate—has stunned users, including academics and some in the tech industry, according to CNN business reporter, Samantha Murphy Kelly. But how effective is it at understanding humour?

In his Substack online platform ('Machine Learning Everything') tech blogger, 'Branko,' wrote:

> *I was a little disappointed because all my other experiences with ChatGPT blew me out of the water. I thought it would be much better than GPT-3 in explaining jokes, but it seems only marginally better. At least it works well with prompting but it tends to just repeat the joke. It doesn't understand word play very well and the ones it does are very common tropes.*

The prospect of AI understanding why we are laughing, and one that can generate its own genuinely funny material, is sort of a holy grail for a subset of AI researchers, according to *Time* magazine's Corinne Purtill:

> *Artificial intelligence can diagnose tumors, read maps and play games, often faster and with more accuracy than humans can. For the moment, however, linguistic humor is still primarily a people thing.*

In my humble opinion, the worldwide rollout of AI is no laughing matter in more ways than one. Besides soulless, transhumanist zealots, who wants to live in a humourless, post-Kafkaesque technocratic dystopia run by robots?

Even the tyrannical bureaucracy in Kafka's nightmarish world had a human face to communicate with, whereas today, in a technocratic age, the machines are void of heart and soul. They even lack a sense of humour.

Think about such a world and its threat to our freedom and privacy: What kind of person wants an under-skin surveillance microchip stuck in his skull in such a dystopian, humourless nightmare, where robots could monitor our behaviour 24/7 and report people who use 'trigger words' or banned phrases in 'offensive' jokes?

But back to AI understanding punchlines. Imagine a robot, a drunk, and a comedian walking into a bar and having a drink together over a few jokes. The chances of the comedian and drunk laughing at the punchlines to the jokes is quite good. But could the AI robot get such punchlines or even understand the jokes in the way they're intended to make someone laugh?

Surely, AI is more intellectually sophisticated than a drunk. However, while the drunk and comedian intuitively get the punchlines, AI has to mine the available data on the internet and form it into intelligible concepts and the selected language spoken by the humans.

It works like this: ChatGPT mimics human speech by absorbing trillions of such words and conversations that it reads online. It absorbs how to associate sentences and phrases, then turn them into concepts, arguments, poems, and jokes.

So, it seems reasonable that syntactically, AI might 'understand' the joke as a basic narrative, but could it understand it semantically or even laugh at it, puns and all? And can it comprehend the method of allegorical interpretation?

As a humorous example, consider the following joke: An old man hobbles up to the ice-cream parlour and orders a chocolate sundae. As he slouches at the counter rubbing his aching upper thigh, the lady asks him what dressings he'd like. She says: "Crushed nuts?" to which the old man replies: "No. Arthritis," as he continues to rub his aching leg.

In order for AI to get the punchline of this joke, it would have to select from over several options: The rude slang term 'nuts' (male testicles, the intended punchline); nuts (a fruit); 'nuts' (slang term for someone who is crazy); 'nut' (person's head); or a nut from a toolkit: a pre-formed metal block with a 'female' internal screw thread.

So, while the drunk and comedian might laugh at the 'nuts/arthritis' punchline, the robot would remain dumbfounded. Now imagine a robocop pulling someone up for speeding down the highway. In an endearing humorous way, the driver says to the robocop: "Ah, come on, officer, I wasn't going *that* fast. I could kill you for pulling me over."

Such a response could be fatal for the driver if the robocop failed to understand the metaphor, took the statement literally, and acted in self-defence. Think I'm joking? ("San Francisco supervisors vote to allow police to deploy robots that kill": Headline in *The Los Angeles Times,* Nov 30, 2022.)

But there's more to misinterpreting a metaphor in the brave new world of AI and transhumanism. Consider The Declaration of Independence of the United States of America:

> *We hold these truths to be self-evident, that all men are created equal, that they are endowed*

> *by their Creator with certain unalienable Rights, that among these are Life, Liberty and the pursuit of Happiness.*

Such a statement might seem immutable and objectively morally sound. Unfortunately, the AI/transhumanist community is not renowned for its deep faith in a transcendent Creator. AI is certainly not conducive to human, spiritual flourishing.

To counter any future threats to our freedom and human worth by AI/Transhumanism, we need to be on guard with such elites who programme the moral philosophy into machines that are more likely to be driven by 'might is right' Darwinian imperatives.

AI/Transhumanism dehumanises us; we lose our charm, worldly innocence of 'the mysterious,' and even sense of humour. The high-tech zealot might argue, 'but it also opens wide new horizons.' Really? To quote a line from Patrick Kavanagh's poem 'Advent': "Through a chink too wide there comes no wonder".

Ultimately, Man alone, or collectively, can never become God in the Transhumanist Revolution, as all revolutions end in failure. But, regarding AI in moderation, and if Man acts theologically through the Logos, he can achieve objective ethical imperatives for the greater good of fellow human beings.

In her 1980 dystopian song "Eighth Day," Hazel O'Connor sums it up well when she echoes the first chapter in Genesis by replacing God with Man. The first line reads: "In the beginning was the world, Man says let there be more light."

The lyrics in the last verse highlight the outcome after seven days of dystopian creation when a problem Man had never anticipated occurs when the "machine just got upset", resulting in a void forever night, where nobody laughs or cries; a world at an end when everyone dies, "forever amen!"

Fortunately, with Logos rising, I believe this science fiction scenario will never happen.

Chapter 3

Can Sewer Pipes Have Minds?

Despite the above headline appearing silly, a similar question, albeit less facetious, was posed by a biologist regarding the sun. The biologist, Rupert Sheldrake, ironically referred to the notion of our main-sequencing star having consciousness (panpsychism). But I want to take it a step further and run with AI and every-day domestic objects, including diapers and sewer pipes—which are seriously doomed if they have minds.

According to some (not all) panpsychists, everything is conscious, even inanimate objects. If that's true, I'm reminded of the old Simon and Garfunkel song, *El Condor Pasa (If I Could)*, when they sang, "I'd rather be a hammer than a nail." If panpsychism is true, I'd rather be a hammer, that's for sure. But it's not just metal nails that suffer, if panpsychism is true. Spare a thought also for the fate of toilet rolls and toilet bowls. And let's not forget diapers, handkerchiefs, cotton buds, and pimple cream.

Am I being puerile or facetious? I stand guilty; it's intentional. But jokes aside, only in academia, hijacked by scientism and anti-theist philosophers, would such nonsense be taken seriously in believing everything possesses consciousness.

When the bizarre topic came up recently, Dr. Rupert Sheldrake asked in the *Journal of Consciousness Studies*: "Is the sun conscious?"[3] It seems Sheldrake has rightly noted a "recent panpsychism turn in philosophy."

This is not surprising as the hard problem of consciousness is the bane of materialist philosophers' frustration ever since the ancient Greek philosophers pondered on the question of the mind and its thoughts.

In the popular magazine *Philosophy Now*, Professor Philip Goff wrote that panpsychism is increasingly being taken seriously in both philosophy and science, but it is still not unknown for panpsychists to receive the odd incredulous stare. "The

---

[3] Rupert Sheldake (2021), "Is the Sun Conscious?" *Journal of Consciousness Studies* 28, No. 3-4, pp. 8-28. Online at https://www.sheldrake.org/files/pdfs/papers/Is_the_Sun_Conscious.pdf

supposition that electrons have some form of consciousness, albeit extremely basic, is still thought by many to be just too crazy to take seriously," he said.

Prominent philosopher David Chalmers, who coined the term the "Hard Problem of Consciousness," has also said: "We're not going to reduce consciousness to something physical ... It's a primitive component of the universe."

Mind Matters online magazine noticed: "But Sheldrake might have added that there is a panpsychist turn in science as well. After all, a mainstream neuroscientist recently argued in a science publication last year that even viruses are intelligent. And he's hardly the only prominent panpsychist in science."

Even *New Scientist*, long a bastion of materialism (naturalism), offers a sympathetic account of panpsychism, which is similar to animism, the ancient belief that everything has a spirit.

This is highly unlikely because mental states are not physical, despite the brain manifesting them via communication with other human beings and/or the external world.

I wrote about this in an essay entitled "The Brain is Not the Mind", where I pointed out that computers and all things non-human are 100% physical. And on the question of AI ever having consciousness: How could a robot ever have a spiritual vision of reality? How could it even be emotional? And even if it could tap into our feelings, how confused would it be with such a complex network of semantics and emotional states?

In fact, it would probably view movies like *Deliverance* as a love story where Cupid's arrow goes straight through the heart of a hillbilly shouting, "Squeal like a pig!" For that matter, could pigs have consciousness? Or sausages? And what about bees?

Science writer at *Mind Matters*, Denyse O'Leary, wrote: "What exactly, does 'consciousness' or 'feel and think' mean when applied to a bee? This usage is no remote outpost. Renowned USC neuroscientist Antonio Damsio tells us that viruses are 'intelligent.' Similarly, University of Chicago biochemist, James Shapiro, tells us in a scholarly paper that all living cells are 'cognitive.'

Could any object at a funeral ceremony feel emotion or be aware of the weeping human mourners?

And if panpsychism is true, would coffins be aware of containing the rearranged atoms of deceased homo sapiens?

Believers in panpsychism claim that not every inanimate object is conscious, but isn't that a bit like wanting your 'panpsychist cake' and 'eating it, too'? Surely, then, any system is conscious, to be consistent with panpsychism. Chalmers speculates: "Rocks will be conscious, spoons will be conscious, the Earth will be conscious. Any kind of aggregation gives you consciousness." If this is true, which I doubt it is, then I'd rather be a forest than a sewer pipe.

But forests, valleys, sewer pipes, and every other existing thing in the universe didn't always exist. This brings us back to creation of the cosmos and the existence of a Necessary Being, and does such an entity *ground* the existence of all existing physical and abstract objects?

Tables, chairs, oceans, and mountains didn't just pop into existence or evolve out of nothing. Such physical things, even if they are to contain consciousness, had to depend on a non-physical, all-powerful, eternal creator. Some philosophers call that a Mind; I call it God. In a nutshell, I believe that the concept of panpsychism is a bizarre solution to the difficult Hard Question of Consciousness that frustrates materialistic philosophers and scientists. They desperately want to fit it into the natural world, even if it takes something spectacularly surreal.

The philosopher of Mind, Ed Feser, says that brains are the most complex things in the universe: "Why suppose that *all* matter, and especially the most elementary matter, is plausibly modeled on them? Surely, the *prima facie* far more plausible bet would be that most matter is radically *unlike* brains."

But it's not just panpsychism that has allegedly come to the rescue of materialist philosophers and scientists. Another surrogate concept to avoid using the 'G' word is the so-called multi-universe, which claims that the fine-tuning of our universe is nothing special in such a gigantic ensemble of other universes. So, while Mount Everest gazes up at the cosmos and marvels at the multiverse, spare a thought for your handkerchief as you blow your nose.

Chapter 4

The Mind is Not the Brain

Is AI on the cusp of reading our thoughts (consciousness)? Can a computer or some form of AI ever achieve this? It is highly unlikely because mental states are not physical, despite the brain manifesting them via communication with other human beings and/or the external world.

Computers and other forms of AI are 100% physical devices, thus the best they can ever achieve at expressing our thoughts is by disrupting them in the 'machinery' of the brain. In other words, frustrating the 'ghost' (consciousness) from operating the 'machine' (wet physical brain).

Think about it: How can a robot ever have a spiritual vision of reality? How can it even be emotional? And even if it could tap into our feelings, how confused would it be with such a complex network of semantics and emotional states? In fact, it would probably view movies like *Deliverance* or *Fatal Attraction* as love stories.

In an article in the *U.S. Sun* newspaper, Ellie Cambridge wrote: "The first ever recording of a dying brain has revealed we might relive some of our best memories in our last moments. Scientists accidentally captured our most complex organ as it shut down, showing an astonishing snapshot into death."

According to the article, the study was published in *Frontiers in Aging Neuroscience*, which claimed their data provided the first evidence from the dying human brain in a non-experimental, real-life acute care clinical setting and advocate that the human brain may possess the capability to generate coordinated activity during the near-death period. This is just one single case study, with a brain that had already been injured due to epilepsy.

The German philosopher Gottfried Wilhelm Leibniz (1646–1716) wrote: "It must be confessed, moreover, that perception, and that which depends on it, are inexplicable by mechanical causes, that is, by figures and motions. And, supposing there were a machine so constructed as to think, feel, and have perception, we could convince of it as enlarged and yet preserving the same proportions, so that we might enter it as a mill. And this granted, we should only find on visiting it, pieces which push one against another, but never anything by which to explain a perception. This must be sought for, therefore, in the simple substance and not the composite or in the machine (Section 17, *Monadology*, 1714)."

In ancient times, the debate on consciousness/thoughts (the mind/body problem) goes back to Plato and ancient Greece. However, it really got kick-started with Descartes and Locke during the 17th century, the former saying, "I think, therefore I am." Could a robot ever think such a thing? Very doubtful.

Without even a basic understanding of what consciousness is, the idea of putting it into a machine, while not difficult to imagine in the fantasy of science fiction, becomes almost impossible to grapple with when it comes down to real and practical implementation.

The field of AI in the last twenty years has made many claims that one day we will create a robot with a consciousness similar, if not the same, as our own. So far, no one has achieved this.

Yet, even though the field of AI has evolved on tasks of solving practical problems such as complex scheduling, rather than on emulating human behaviour, many AI scientists still believe that the original goals of AI will become a reality in the near future. Unfortunately, a lot of scientists don't believe in the human soul.

But even without recourse to the notion of a soul, there are several ways in which the mind is not simply brain. The wet, grey organ known as the brain does not have mental states such as love, hate, or sadness. Then there is the problem of propositional attitudes such as fear, hope, desire, wish, dread, and regret.

As to where the mind resides, that is the biggest mystery in philosophy. Although it interacts with the brain, it can't be a kind of invisible vapour hovering above one's head. Nor can it be located in some part of the universe, as it's supernatural and outside of space and the material world.

If epiphenomenalism (mind is brain) is true, then how come fake drugs sometimes work in placebo effects? And at what point in evolution did the atoms in brains develop morals? That we can have logic, reason, and truth evolving out of a material process that is aimless, purposeless, misguided, and unaware of self seems absurd.

Materialism would never have endowed human beings with consciousness, as zombie clones would reproduce and spread their genes more effectively. If intelligence emerged from brute matter and not a Superior Mind, then robotic ideas of life and the cosmos are not to be taken seriously because they could never tackle metaphysical questions reliably. Should one trust a machine made of blind chemicals or computer software functioning on syntactic information but oblivious to philosophical truths?

For if a robot could read thoughts as material entities that would also have to include laws of logic, which is absurd. One can hit a rock or a tree with a hammer but not a law of logic. The same applies to the existence of objective moral values and duties, aesthetic values, love, the existence of minds other than one's own, the duration of the past, beauty, science, and the existence of the external world.

Could a carbon android at the funeral of a child read the thoughts and emotions of the mourners? Or would it view the death of a loved one as nothing more than the rearrangement of atoms in a wooden box?

Some scientists speak as though the mysteries of consciousness and mental experience have been fathomed. That is certainly not the case. There is something non-material about thought and experience which has not been explained by scientific materialism.

To draw an analogy: the brain is like a motor car, and the mind a human who drives it. If the car ('brain') is damaged or the engine not tuned properly, then the driver ('mind') will not be able to drive it properly. So, rest assured: The only one who can ever read your mind is you.

Chapter 5

Beauty and the Curse of Relentless Efficiency

Is it just me or is almost everything in the seemingly dying embers of Western civilisation really soul-destroying? Think about it: From hi-technology (dentistry excluded!), modern architecture to movies, songs, literature (non-fiction excluded), fashion, or all the other arts and crafts, the aesthetics of anti-culture over the past 20 years sucks to high heaven. And, although it's been brewing since the late-1960s, it seems to have peaked at the beginning of the 21st century and is linked to an obsession with relentless efficiency.

When I hear what passes for music on the radio or stroll through the city streets and gaze at buildings or a theatre poster, I wonder how in the name of God did we go from the melodies of Beethoven and Bach to Hip-hop Rap; from Greco-Roman architecture to Brutalism, and from Shakespeare's *Hamlet* to the blandness of what passes for theatre today, despite the odd good play.

Regarding Brutalism: I don't have a problem if the structure is in an appropriate space and not in an area where classic architecture exists, as the juxtaposition of such buildings can spoil the view of a great cathedral or city hall. Theodore Dalrymple said he once asked an architectural historian why it was that, after so many hundreds of years of architectural achievement, we in Europe were all but incapable of building an aesthetically decent house, let alone an elegant and achieved public building. "He did not deny the premise of my question," said Dr Dalrymple, "but replied that it was because of modern techniques of building and the materials architects had to use (no choice in the matter). In other words, it was a purely technical matter."

Dr. Dalrymple added that he did not believe it then and does not believe it now. "But I do believe that there are economic and quasi-economic reasons for building as we build 'cheaper, faster, easier, larger and so forth,' but the irresistible force of these arguments in practice indicates our scale of values. It is not so much that we cannot, it is that we will not." (*Takimag*, Sept 07, 2014)

There seems to be no end to the dumbing down of what once rich in aesthetics of Western civilization, despite its many flaws in a fallen world. As for efficiency: We seem to be reaching for an efficiency only to strive to become more efficient, thus ending up as slaves of technological machines or devices, as well as philistine vandals of high art.

Some of what we are witnessing today was forewarned thousands of years ago in the Bible: 'Woe to those who call evil good and good evil, who put darkness for light and light for darkness, who put bitter for sweet and sweet for bitter' (Isaiah 5:20).

And in Colossians 3:23, St. Paul said: 'Whatever you do, work at it with all your heart, as working for the Lord, not for human masters.' Paul is saying that if you work through the Lord, the outcome will be that derived of the Logos: Beauty, Logic, etc, whereas working for human masters is open to error or aesthetic darkness similar to much of today's low culture. In Titus 1:15: 'To the pure, all things are pure, but to the defiled and unbelieving, nothing is pure; but both their minds and their consciences are defiled.'

For the German philosopher, Martin Heidegger, who died in 1976, such things, especially technology, can deprive us of our essence as human beings. As well as the efficiency of technology being a threat to our humanity, there are other fields of life where efficiency clashes with beauty. A good example of this efficiency verses aesthetics was given by Hubert Dreyfus during the 1980s BBC TV series *The Great Philosophers*, hosted by the late Bryan Magee. In the programme featuring Heidegger's philosophy, he said: "We don't even seek truth anymore but simply efficiency. For us, everything is to be made as flexible as possible so as to be used as efficiency as possible." He said a good example of this was a Styrofoam cup: it keeps hot things hot and cold things cold, and you can dispose of it when you are done with it. It efficiently and flexibly satisfies our desires.

He added: "It's utterly different from, say, a Japanese teacup, which is delicate, traditional, and socialises people. It doesn't keep the tea hot for long, and probably doesn't satisfy anybody's desires but that's not important."

Likewise, in modes of transport, a high-tech, efficient, superfast, sound-proof train is not as pleasurable to travel in compared to an old, slower locomotive with its quaint interior carriages and the sound of the tracks gently 'banging' as the steam engine occasionally blows its horn on approach to level crossings. For us humans, as we travel through the third decade of the 21st century, we've reached a crossroads where, as CS Lewis puts it, we might have to do an about turn and walk back to the right road in order to be the most progressive human.

Chapter 6

Is Berkeley's Universe a Spiritual Hologram?

Ronald Arbuthnott Knox, an English clergyman and writer, had his claim to fame in a much-forgotten BBC radio hoax in 1926. The broadcast, which pre-dated HG Wells's more famous 1938 alien war hoax, was a simulated live report on a revolution taking place in London. But this essay is not about Wells nor Knox; it's about someone else's work which Knox mocked:

There was a young man who said God
Must find it exceedingly odd
To think that the tree
Should continue to be
When there's no one around in the quad.

An anonymous reply to Knox's limerick added:

Dear Sir, your astonishment's odd;
I am always about in the quad
And that's why the tree
Will continue to be
Since observed by yours faithfully, God.

That 'tree that fell in the forest' did make a sound after all, as witnessed by God. As Berkeley famously said, "To be is to be perceived." Born in County Kilkenny, Ireland, on 12th March 1685, Berkeley studied at Trinity College, Dublin. He became a Fellow in 1707, and this required him to be placed in clerical orders, thus leading to his ordination in the Anglican Church.

Berkeley moved to America in 1728 and tried to found a missionary college in Bermuda. He abandoned this plan in 1732, but he had a great effect on higher education while in America, assisting in the development of the Columbia universities, Yale and several other schools (what would he make of these institutions today?). In 1734, he was made Bishop of Cloyne, Ireland, where he remained. He died on January 14, 1753.

In 1710, he published *The Principles of Human Knowledge,* but it failed to win over readers to his immaterial theory. He then published a more popular version, *The Three Dialogues Between Hylas and Philonous,* in 1713. Much of his philosophy was regarded as silly by other philosophers during this period, but they couldn't refute it. So, what to make of the *Dialogues*? With apologies to Berkeley, below is an edited, dumbed-down, modern-day version that I wrote of what Rock (Hylas) and Barkley (Philonous) had to say. Picture the scene: a quiet bar somewhere in the West. Rock is nursing a pint, as his buddy Barkley walks into the pub.

Rock: Well, if it isn't the philosopher . . . or just a thought.

Barkley (sits beside Rock): Are you buying me a drink? And what's this, "just a thought" remark?

Rock: Last night, the lads were having a great laugh at your whacky ideas on ideas.

Barkley: There's nothing whacky about them. Now, where's my drink?

Rock: Well, if you're just a thought, then I won't have to buy you a drink. In fact, according to you, the drink doesn't exist—or me for that matter . . . hic!

Barkley: You think it's silly? What if I prove that your way of thinking is nonsense and riddled with contradictions?

Rock: Go on then, persuade me. And if you do, I'll buy you that drink.

Barkley: I'll buy my own. You sound a bit tipsy. How many pints did you skull?

Rock: Four . . . five. I can't remember . . . hic!

Barkley: And the drink is relaxing you?

Rock: That's true.

Barkley: Making you more sociable?

Rock: It's good to see you, Barkley.

Barkley: You feel a little dizzy.

Rock: Just a little.

Barkley: And a bit sentimental.

Rock: I miss my mother.

Barkley: Tell me, Rock. If you drank more, would those feelings intensify?

Rock: That's for sure.

Barkley: And is anything that can't be perceived capable of pleasure or pain?

Rock: Of course not . . . hic!

Barkley: Is your pint of stout a senseless being or a being endowed with sense and perception?

Rock (taking a slug): Ahh . . . It's senseless, but it tastes great.

Barkley: And it's cool?

Rock: Chilled to perfection.

Barkley: A chilled liquid?

Rock: It certainly wets my whistle.

Barkley: It cannot therefore be a subject possessing relaxation, sociability, dizziness, sentimentality, coolness, and liquidity.

Rock: I agree.

Barkley: Does it not then follow that without a mind to perceive the pint, the pint and its properties then cease to exist?

Rock: Is this like a pint falling off a tree in the forest and not making a sound?

Barkley: Kind of. Only such a pint would make a sound.

Rock: How, if it's not perceived?

Barkley: Everything's perceived in the mind of God; and, in turn, that pint you're drinking is dependent on your tiny mind for its existence. Now buy me that drink.

According to Berkeley, Rock's visible, tangible pint is caused by God. For Berkeley, God or religion is the basis for improving one's life, not damaging it, and a common-sense view of life is the best path to achieving such a goal. Berkeley was unique and quite happy in his philosophical worldview while he thought other philosophers throughout history were usually frustrated and complicated matters by analysing everything beyond the reach of human reasoning. Being reasonable for Berkeley is being anti-sceptical and acknowledging that the only 'real' things that exist in the world are spirits who are created by an infinite Spirit, God. Put simply, the whole of reality is mental (certainly in the 21st century, in more ways than one).

Now at this point, one might argue that this so-called 'real world' does not deserve the title of reality. But for Berkeley it is, to a degree, in that it is a kind of second-class reality always one step in the shadows of God's Mind and the minds of finite beings. What we perceive as matter plays second fiddle to Spirit. This philosophical theory was developed as an answer to scepticism and atheism that had crept into contemporary philosophy.

Berkeley hated atheism and wanted to put God centre stage, acknowledging that an Absolute Observer must reign supreme over perceiving reality. Such a supernatural eternal Absolute Entity would not require the multiplication of causes beyond what is necessary to explain itself.

However, human ideas/consciousness require an explanation for its meaning as the alternative of solipsism (the belief that only oneself exists and everything else is an illusion) is less credible for explaining one's existence. So, we are left with the ultimate question: is consciousness/ideas connected to and part of God's Mind?

Are the words that you are now reading on this page, including the backdrop to wherever you are reading, part of the conceptualized reality of a Supreme Entity, of which our collective consciousness is a manifestation? Berkeley believes that everyday objects are such a manifestation with multiple visual aspects that can change depending upon the circumstances. This also brings into play the problem of appearance and reality.

Take for example a fingerprint. If asked to describe one, the obvious answer would be that it's a small black blob, about two inches in circumference, with whirly lines going through it. The philosopher Bertrand Russell would call this favouritism, as we tend to view objects from an ordinary point of view under usual conditions of light, but the other colours/shapes which appear under other conditions have just as good a right to be considered real.

In other words, if we look a little closer, through a powerful microscope, our conventional idea of what a fingerprint looks like takes on a whole new meaning. For here we see something that resembles a huge mountain range, a kind of dark grey version of the Himalayas.

And if we stand back from the fingerprint, say about 20 feet, it looks like a tiny black spot (without the whirly bits). The same applies to everything else we perceive—from tables and chairs to mountains and oceans. But we describe most everyday objects from the very convenient distance, usually a couple of feet away, of a human perceiver, with its meaning relative to how such a perceiver thinks.

Can we know what reality looks like if everything is observer relative? This appearance-and-reality question has dogged philosophers for thousands of years whereas scientists don't waste time with endless speculation and abstract analysis. In recent years, many physicists have claimed to provide "some of the clearest evidence yet" that our universe could be just one big projection: a hologram. But that leads to another question: who is operating the hologram?

On Berkeley's Idealism, scientists could be accused of philosophically putting the cosmological cart before the metaphysical horse, i.e. the so-called 'world' of phenomena before initial experience of it. Some scientists would argue that the tree fell and made a sound because it is part of the material world, which exists independently of our consciousness. In other words, it is mind-independent. The same, they would argue, applies to the universe and all the heavenly bodies it contains. But then most scientists are not philosophers like Berkeley. Berkeley would have probably thought that scientists inadvertently invent the rules in order for us to play the 'material' game of life.

But this can be a good thing, in that it gives us a kind of practical structure to our day-to-day living. Ultimately, it tells us nothing about consciousness or why there is a weird, vast configuration in the form of what we call the universe. Or is it a God-constructed hologram, not the scientific atheistic view, a kind of cosmic window-dressing to keep us in awe of the astonishing vastness of the cosmos?

Berkeley's philosophical theory was developed in contrast to the ideas of British philosopher John Locke (1632-1704). Locke made the distinction between ideas, immediately aware in perception to people, and material things caused by such ideas. This distinction runs counter to Berkeley's philosophy in that ideas are the first (most immediate) things that spark an awareness in human beings, as in sound, touch, sight, smell, and so forth.

Locke was trying to find out what is it that is out there; Berkeley looks from within as his starting point. Chairs, tables, trees, and everything in the so-called material world are not active in the same way as minds are. Berkeley is confident in the knowledge that such objects, unlike minds, are not perceiving and do not have ideas (panpsychism). And everything is in the infinite Mind of God.

In *Of The Principles Of Human Understanding*, he wrote: "It is indeed an opinion strangely prevailing among men that houses, mountains, rivers and in a word, all sensible objects, have an existence, natural or real, distinct from their being perceived by the understanding . . . For what are the aforementioned objects but the things we perceive by sense? And what do we perceive besides our own ideas or sensations? And is it plainly repugnant that anyone of these, or any combination of them, should exist unperceived."

But what if *your* perception is all there is—without God? As mentioned, this is called solipsism, meaning reality is an illusion and only you exist reading these 'imaginary' words, which you think were written by someone else. Your mind, and your mind alone, is just one uncanny brutal fact. Other 'people' are just phantoms in your imagination. This seemingly irrefutable argument can be rejected by a leap of faith based on the understanding that only a mad person would believe in it. However, Berkeley's answer to solipsism is that God can perceive sensations independently of us, thus allowing "things" to exist while we are not actually looking at them.

But what if God is not the good God that Berkeley believes in but an evil demon deceiving *you*? Such a demon, right now, is giving you a hint of his existence by conjuring up the text that you are reading. Even the 'fictitious' Berkeley has been invented to deceive you for the amusement of this evil demon. This hypothesis, from French philosopher Rene Descartes (1596-1650), throws doubt on the evidence of your senses. The demon has created the astonishing illusion which is *your* world. This is similar to the popular movie 'The Matrix'. In it, reality perceived by the humans is really the Matrix: a simulated reality created by sentient computers.

But Descartes notes that even if such an evil genius existed, there is one thing that he, Descartes, knows for sure: "I think, therefore I am." And he believes he cannot be deceived of this. Berkeley would probably phrase it another way: "I think; therefore, God exists."

The Greek philosopher Plato (circa 428-c.348) would attribute his existence and thinking on a higher form. Chairs, tables, mountains, oceans are reflections of the real objects, which he called the Forms.

He gives this analogy of a cave where some slaves are shackled in chains. These slaves see a series of shadows on the cave wall, passing back and forth, which they take to be reality. One of the prisoners is released and brought up near the cave's entrance, where he sees a fire and, in front of the fire, people are walking around carrying statues, animals, everyday objects. The prisoner, rubbing his eyes because of the great light caused by the fire, thinks to himself: "I've been misled." Up to that point, he had been seduced by what he immediately saw before him. The shadows where just representations of the higher order of forms.

There is no doubt that Berkeley views God as the Absolute Form with the highest order of mind, and the everyday 'objects' making up the furniture in the so-called universe as immaterial reflections.

In *The Principles of Human Knowledge*, Berkeley writes:

> . . . If he can conceive it possible either for his ideas or their archetypes to exist without being perceived, then I give up the cause; but if he cannot, he will acknowledge it is unreasonable for him to stand up in defence of he knows not what, and pretend to charge on me as an absurdity the not assenting to those propositions which at bottom have no meaning in them.
>
> It will not be amiss to observe how far the received principles of philosophy are themselves chargeable with those pretended absurdities. It is thought strangely absurd that upon closing my eyelids all the visible objects around me should be reduced to nothing; and yet is it not this what philosophers commonly acknowledge, when they agree on all hands that light and colours, which alone are the proper and immediate objects of sight, are

> mere sensations that exist no longer than they are perceived . . . ?

For Locke, colours are relative to the minds of perceivers. He called them secondary qualities, along with smells and tastes because they were 'mind-dependent'. However, the shape of an orange is a primary quality. And such material objects, according to Locke, are fully mind-independent regardless of how they look to us. Berkeley does not agree. He points to subjective experiences such as the sensation of pain. Would the experience of pain exist somewhere out there in mind-independent land? So, for Berkeley, Locke's primary and secondary qualities do not make sense.

But back to colours. The Australian philosopher Frank Jackson highlights what it is like to see a colour, as opposed to its so-called physical or structural properties. Imagine a woman called Mary who is an expert on human vision, particularly colour perception. Mary has never seen a colour and has spent all her life working in a black-and-white laboratory with some grey areas. Even her TV is black and white. She knows what happens in humans' brains when they see a colour, as she spent her life reading about light waves, particles, and optical perception.

One day, feeling extremely bored, she walks outside the lab and sees a front yard with green grass. Mary learns something new: what it is like to see something green. All the science on colour optics in all the books she read are nothing like what she has just experienced. All the hypothetical information on physical and structural properties of colour experience that she learned in the lab have been blown away. She has, for the first time, learned about the conscious aspect of seeing green. And that conscious aspect appears to be non-physical.

At the end of the Moody Blues' song 'Nights In White Satin' (extended version), a short poem called 'Late Lament' is narrated which touches on the mind-dependent colour issue. Here is the last verse:

> *'Cold hearted orb that rules the night,*
> *Removes the colours from our sight,*
> *Red is grey and yellow white,*
> *But we decide which is right.*
> *And which is an illusion.'*

But life is not all an illusion for Berkeley, as he counts his ideas as real. Even the atheist British philosopher John McTaggart (1866-1925) rejects the existence of God while acknowledging the spiritual nature of reality. This is what he had to say about Berkeley:

> Our method will not be epistemological. We shall not start from the consideration of beliefs only. We shall, on the contrary, endeavour to determine those general characteristics which apply to existence as a whole, or to everything that exists, whether these things are beliefs or not.
>
> And this, while the result which we shall reach will prove to be the one which would be usually, and properly, called Idealism, it will be the Idealism of Berkeley, of Leibniz, and—as I believe—of Hegel.
>
> It will not be the Idealism of Kant, or of the school which is sometimes called Neo-Hegelian. It will not, that is, be the Idealism which rests on the dependence of the object of knowledge upon the knowing subject, or

> upon the fact of knowledge, but the Idealism which rests on the assertion that nothing exists but Spirit.

German philosopher Immanuel Kant (1724-1804) would certainly have a problem with nothing existing, only Spirit. Berkeley's Idealism does not fit well with his philosophy. For Kant, 'God' exists, but not Berkeley's God.

He argues that it is impossible to prove God exists, but he nonetheless believes He exists through faith. But McTaggart's mention of German philosopher Georg Hegel (1770-1831) is interesting. Hegel's ideas are notoriously difficult to understand. He seems to reject the subjective form of Idealism in favour of universal mind. And this idea of the universe as one mind is no stranger to some Eastern philosophies: one reality; one mind.

But Berkeley was a Christian, and his Idealism encompasses other 'sub' minds, a kind of community of souls in a non-physical realm. In the end, hologram or no hologram, speculation or scepticism on the universe and the world are a waste of time. Life is short and to lose your eternal soul sitting on the fence of agnosticism is worse than tragic: it's the stuff of nightmares if God exists.

Hebrews 11:3 tells us, 'By faith we understand that the universe was created by the word of God, so that what is seen was not made out of things that are visible.' A faith in God based on reason and trust (not blind faith) is all we need for physical, spiritual fulfilment and truth. The truth is more important than a worldview you *want* to live with based on a false perception that is 'my reality.'

Now, where did I put that bottle of Hologram Cabernet Sauvignon Fortuna? God only knows where I left it last.

Chapter 7

Genesis, a Cup of Tea, and the End of the World

From a Christian perspective, in September 2017, in an epic hit-and-miss essay in *The Atlantic* on America losing its mind, Kurt Andersen writes that Americans believe—"*really believe*"—in a story of life's instantaneous creation several thousand years ago.

If Mr. Andersen is referring to the age of the universe or life, he is exaggerating. All Americans don't '*really believe*', although some do, in instantaneous creation several thousand years ago. From the Belgium Catholic priest, astronomer Georges Lemaitre (1894-1966), to astrophysicist Stephen Hawking, some 13.7 billion years seems to be the duration in time since the universe began, according to Big Bang cosmology. And most American Christians and atheists—again, not all—believe in this and an old universe/earth. As for life: it didn't just pop out of nothing *uncaused*, a concept that is religiously neutral.

However, Christianity, which is in fact the most science-friendly belief system amongst the world's religions, says nothing about the earth's age. As for human creation: The story of primordial beings, Adam and Eve, is infused with metaphor and symbolism. The Hebrew for Adam is 'Man', and Eve is 'Living One'. There was no talking snake, but a metaphor symbolizing Satan (serpent). However, this is an inhouse discussion amongst fellow Christians and the subject of another essay. But back to creation: Stephen Hawking, although most of his views on astronomy are sound, is dismissive of the biblical account. He also rejects a creator.

About seven years ago, during a talk on Hawking at a university, I raised my hand and criticised comments he made in his then latest book *The Grand Design*, which he co-wrote with Star Trek screenwriter Leonard Mlodinow.

My question was, "Why did Hawking write such a nonsensical idea that the universe created itself because of gravity?" (In order for the universe to create itself it would have to have existed before it exists, and gravity is part of the universe). I also asked why did Hawking write "philosophy is dead" at the beginning of his book (a self-refuting statement, as it's philosophical) while also constantly philosophising throughout the entire book?

There was an awkward silence in the lecture hall, and the speaker looked at me in what seemed like a confused expression. He said, "Did he really say that?" (He hadn't read the entire book). I told him the page numbers where he could find the quotes. I wasn't criticising Hawking the man (a man enduring a severe neurone disease that has paralysed him for decades), but Hawking the scientist.

But, as the speaker looked at me with what seemed like an expression of disbelief, to my rescue came a distinguished astrophysicist on the panel, who stood up and said, "Kenneth is right; Hawking did write those things". The subject was quickly changed in a 'move-along-nothing-to-see-here' kind of way. Because Hawking is an atheist and doesn't believe the universe was created by God, his views on the origin of the cosmos generally go unchallenged by secular academics and the mainstream media who are, for the most part, hostile to Christianity.

And he's not alone in this view. The likeable atheist philosopher Daniel Dennett also denies that the universe has any theological importance. Although he says that the universe has a cause, he thinks its cause is: 'Itself' (*WHAT?*). Yes, he calls this the 'ultimate bootstrapping trick' (is this a non-existing universe, sometime in the finite past, pulling up a non-existence 'itself' by non-existence bootstrings? Where do these metaphorical bootstrings come from? Not to mention 'Itself'!) Not only is this logically incoherent but, in the words of Oxford University's mathematician, John Lennox, nonsense remains nonsense regardless of who says it. Nonsense, because in both Dennett's and Hawking's descriptions, a first state of the universe is beyond scientific explanation: it is outside of Naturalism, where there are no physical laws; therefore, the cause must be an eternal, uncaused, immaterial, personal, all-powerful, and ultimately divine entity, i.e. a non-physical Mind: God.

Then there's the late celebrity scientist Carl Sagan, another likeable atheist, who also didn't accommodate the divine or a creator. He was also no stranger to contradictions and the odd, daft statement about the origin of the cosmos. He once said: "One of the great commandments of science is,

'Mistrust arguments from authority'." (Scientists, being primates, and thus given to dominance hierarchies, of course do not always follow this commandment.)

What he seems to be saying is, 'Don't trust what I'm saying', as he and all the other scientists are hairless apes prone to error. He added: "The cosmos is all that is, or ever was, or ever will be." *Really?* Whatever happened to the Big Bang and the Second Law of Thermodynamics (SLT)?

But the secular academia and 'intelligentsia' would rather believe celebrity scientists than the Bible's account of creation. This rejection of Genesis is where 'intellectual' pride gets you. In 2 Corinthians 11:19, St. Paul said: "For ye suffer fools gladly, seeing ye yourselves are wise." But not all non-believers are hostile to Scripture.

In his book *The Genesis Enigma*, Oxford evolutionary biologist Andrew Parker wonders how the writer of the first chapter of Genesis got it so scientifically correct. Parker is taken aback by the order of creation described in Genesis, which follows the order of geologic and life evolution as science understands it. He writes: "Either the writer of the creation account of Genesis 1 was directed by divine intervention or he made a lucky guess."

Parker isn't alone. Other agnostics, theists, and atheists marvel at the fine-tuning of the universe and its beauty. In his book *The Universe: Past and Present Reflections*, British astrophysicist Fred Hoyle wrote: "A common sense interpretation of the facts suggests that a super-intellect has monkeyed with physics, as well as with chemistry and biology, and that there are no blind forces worth speaking about in nature. The numbers one calculates from the facts seem to me so overwhelming as to put this conclusion almost beyond question."

The opening line of Genesis is historically the first to mention the Big Bang by another name: "In the beginning . . . " For the ancient Greeks and Eastern religions, the universe was/is eternal, but the Bible proved otherwise when a hint of the SLT debunked such unscientific notions. This law began to emerge again in the early 19th century and became widespread and fully developed in the 20th century, estimating the age of the universe at approximately 13.7 billion years old.

But it's not just the beginning of the universe that the Bible got spectacularly right and went against general accepted teaching of the time. For the Bible, there was never a flat earth. *Isaiah 40:22* mentions "circle of the earth," and in *Job 26:10*, God inscribed a circle on the surface of the waters. And what supports the earth? According to the Bible, it's not an infinite number of turtles all the way down, but space. God "hangs the earth on nothing" *(Job 26:7)*. For the earth to be unsupported would have been regarded as absurd in the ancient world. Then there is the expansion of the universe. According to the Bible, the universe has been "stretched out" *(Isaiah 40:22)*. Just imagine how ridiculous this sounded to our ancient ancestors that the seemingly 'static' heavens that they saw in the night sky was in fact expanding.

There is much more on how astronomy confirms the Bible, but back to thermodynamics. To the ordinary person outside of the fields of science, the SLT might seem complex. However, the great philosopher, JP Moreland, gives a good 'coffee' analogy. But to borrow from JP, here's my 'tea' version of SLT: Imagine boiling a kettle to make a cup of tea. You pour the boiling water into a cup, place it on the table, and walk away. Twenty minutes later, you return to the cup and place your finger into the water, which is no longer hot but cooling down until it reaches the point of chilled. This is what is happening on a cosmic scale to the universe: everything is winding down to a uniform state; the lights and heat turned off resulting in heat death of the cosmos: from Big Bang to a pathetic whimper (you'll never look at a cup of tea in the same way again!). The philosophy writer PJ Zwart describes such a cosmic state:

> . . . according to the second law the whole universe must eventually reach a state of maximum entropy. It will then be in thermodynamical equilibrium; everywhere the situation will be exactly the same, with the same composition, the same temperature, the same pressure etc., etc. There will be no objects any more, but the universe will consist of one vast gas of uniform composition. Because it is in complete equilibrium, absolutely nothing will happen anymore. The only way in which a process can begin in a system in equilibrium is through an action from the outside, but an

> action from the outside is of course impossible if the system in question is the whole universe. So in this future state of maximal entropy, the universe would be in absolute rest and complete darkness, and nothing could disturb the dead silence.

This is a major headache for atheists, as the beginning of the universe demands a first cause; in other words, a Creator, while the end of the universe, if there is no God, is cosmic 'curtains' with everything everyone ever did throughout history, *ultimately* becoming undone and in vain. In his book *A Free Man's Worship*, atheist philosopher Bertrand Russell bleakly laments:

> That man is the product of causes which had no prevision of the end they were achieving; that his origin, his growth, his hopes and fears, his loves and his beliefs, are but the outcome of accidental collocations of atoms; that no fire, no heroism, no intensity of thought and feeling, can preserve an individual life beyond the grave; that all the labours of the

> ages, all the devotion, all the inspiration, all the noonday brightness of human genius, are destined to extinction in the vast death of the solar system, and that the whole temple of Man's achievement must inevitably be buried beneath the debris of a universe in ruins—all these things, if not quite beyond dispute, are yet so nearly certain, that no philosophy which rejects them can hope to stand. Only within the scaffolding of these truths, only on the firm foundation of unyielding despair, can the soul's habitation henceforth be safely built.

Russell, who criticised people who were certain of their beliefs, was certain of this depressing outcome. But one must commend his profound command of the ramifications of a Godless universe. But there is a God because something caused the universe, and science points to a First Cause.

Even the world's most famous atheist, Richard Dawkins, gets confused on this question. During a TV debate in Australia a few years ago on the origins of the universe, Dawkins claimed something can come from nothing. He said: "Of course common sense doesn't allow you to get something from nothing . . . SOMETHING [yes, you read that right] pretty mysterious had to give rise to the origin of the universe." Dawkins also said supernatural claims for the universe should be "ridiculed with contempt." Does that include the Big Bang, which was a supernatural event? Or freedom of the will? Or the laws of logic, aesthetic judgments, love, morality, mathematics, all of which transcend Naturalism?

Consider the philosopher William Lane Craig's version of the Kalam Cosmological Argument. Here is the crux of the argument: whatever begins to exist, has a cause of its existence. The universe began to exist. Therefore, the universe had a cause of its existence. And this cause of the universe must be a *personal* creator.

Dr. Craig says the only way to have an eternal cause but a temporal effect would seem to be if the cause is a personal agent who freely chooses to create the universe. Think about it: there was no 'before' the universe because time itself was created with space and matter.

The term, mentioned earlier above, is 'ontologically prior' to the universe. Very few people, with the exception of a handful of philosophers, find God's relationship with time and divine eternity to be enormously complex. It makes the problem of evil and suffering (theodicy), the greatest problem for most people, seem like a walk in the park. Think about it: God has always existed. However, he entered time a finite time ago by an act of making a personal, free choice. So, the hard question is, what 'was' God doing ontologically prior to the universe? Was he floating around in some transcendent dimension whistling and twiddling his thumbs while waiting to create the Big Bang? Such a state seems absurd.

According to Christianity, God is a triune entity (three persons in the one Godhead) and, as such, there is spiritual love relationship within this entity. If God isn't triune, then His love would have to be self-absorbed love, which is an imperfection, just like a narcissist is deeply flawed. And as God is not some finite, human figure, we cannot know exactly what He was doing or thinking ontologically prior to the universe. As for Time: God also knows the future. This is not to say that we are determined and lack free will. Our actions and potential future actions are decided by us and us alone.

As for how the world will end: for the theist, spiritual eschatology is the apocalyptic viewpoint of the Second Coming of Christ. Here is how that event is described in the *Apocalypse of John*, the last book in the New Testament (remember, the Bible is rich in metaphor and symbolism. The NT text below is translated from Greek):

> Then I saw a great white throne and him who was seated on it. From his presence earth and sky fled away, and no place was found for them. And I saw the dead, great and small, standing before the throne, and books were opened. Then another book was opened, which is the book of life. And the dead were judged by what was written in the books, according to what they had done. And the sea gave up the dead who were in it, Death and Hades gave up the dead who were in them, and they were judged, each one of them, according to what they had done. Then Death and Hades were thrown into the lake of fire.

> This is the second death, the lake of fire. And if anyone's name was not found written in the book of life, he was thrown into the lake of fire.
>
> Then, I saw a new heaven and a new earth, for the first heaven and the first earth had passed away, and the sea was no more. And I saw the holy city, new Jerusalem, coming down out of heaven from God, prepared as a bride adorned for her husband. And I heard a loud voice from the throne saying, "Behold, the dwelling place of God is with man. He will dwell with them, and they will be his people, and God himself will be with them as their God. He will wipe away every tear from their eyes, and death shall be no more, neither shall there be mourning nor crying nor pain anymore, for the former things have passed away." —Rev. 20.11-21.3 ESV

Eschatology, says Dr. Craig, has also become a branch of physics called: Physical eschatology. It is a sub-discipline of cosmology, which is the study of the large-scale structure and evolution of the universe. Cosmology subdivides into two parts: Cosmogony is the sub-discipline which studies the origin and past history of the universe. Eschatology, by contrast, is the sub-discipline which explores the future and final fate of the universe.

William Lane Craig again:

> Just as physical cosmogony looks back in time to retrodict the history of the cosmos based on traces of the past and the laws of nature, so physical eschatology looks forward in time to predict the future of the cosmos based on present conditions and laws of nature. The challenge for those interested in the interface between theology and science is how to arrive at an integrated perspective on the world's future adequate to the concerns of both theology and science.

The key to physical eschatology is the Second Law of Thermodynamics. About the middle of the nineteenth century, several physicists sought to formulate a scientific law that would bring under a general rule all the various irreversible processes encountered in the world. The result of their efforts is now known as the Second Law of Thermodynamics.

With all that in mind, why not make yourself a nice cup of tea and don't worry about the end of the world . . . for now.

Chapter 8

Into Huxley's Brave New World

On November 22, 1963, three of the world's highly rated, influential idealists, writers Aldous Huxley, C.S. Lewis, and the American president John F. Kennedy, suffered death. This coincidental triune bereavement kept the obituary writers busy during that fateful day, but Kennedy's death would eventually get top billing, his bloody assassination eclipsing Lewis and Huxley in terms of human-interest appeal.

What's intriguing about these three men is their deep interest in the fate of humankind: Kennedy was suspicious of shadowy figures negatively interfering in world events; Lewis was concerned about the threat of Secularism and disbelief in God; Huxley wrote about the future of a *Brave New World*. This dystopian novel, published in 1932, is set in a futuristic New World Order (based in London), whose population is controlled by the State (sounds like Planet Earth 2026).

Such a dystopia would've been a nightmare for Lewis and Kennedy, but for Huxley, any scenario void of the God of Christianity and ruled by secular oligarchs, was closer to utopia (he did not want there to be a god).

In this *Brave New World,* we see genetically modified citizens born in artificial wombs in a world where the ruling class is superior to the Great Unwashed; a dark technological landscape where the leaders play God and disobedience to the State is forbidden.

Other things forbidden are great works of literature or beautiful, classical art. Such aspects of high culture are banned because they're timeless and deeply appealing to people, who are now brainwashed to consume new things; besides, in the New World State, citizens wouldn't understand or appreciate the great artistic works of the past. As long as they are infantile, 'happy' and zombie-like, their life of slavish bondage will lie hidden and way beyond their shallow intellectual understanding. A people whose spiritual complacency, moral stupor, and lemming-like behaviour resembles that of H.G. Wells's docile Eloi (*The Time Machine*).

Huxley wrote: "A really efficient totalitarian state would be one in which the all-powerful executive of political bosses and their army of managers control a population of slaves who do not have to be coerced, because they love their servitude." In other words, the citizens have deified the State as their God. (Isn't it strange that the rock'n'roll rebels and anti-Establishment heroes of yesteryear are now replaced with Establishment-worshiping celebrities, 'musicians', 'actors' and students? It seems today's leftists just love their ruling-class masters, even to the extent of being their proxy warriors in using identity politics to distract from the economy, poverty, and crime.)

As for the inhabitants of *Brave New World*: For many, technology has replaced God. In the last decades of his life, the existentialist German philosopher Martin Heidegger (1889-1976) spoke about the threat of technology, as opposed to its benefits. In his book *How to Read Heidegger*, philosophy professor Mark Wrathall wrote: "His preoccupation with 'the technological mode of revealing' was driven by the belief that if we come to experience everything as a mere resource, our ability to lead worthwhile lives will be put at risk. His task as a thinker was to awaken us to the danger of this age, and to point out possible ways for us to avoid the snares of the technological age."

It's possible that later on, movies, songs, and events like Kubrick's *2001: A Space Odyssey*, the song *In The Year 2525,* and the Apollo moon landings convinced Heidegger more and more of the power and potential threat posed by technology and what it is to be human. In this 'Brave New World', efficiency trumps truth, and to be real is to be used as efficiently as possible.

Heidegger scholar Hubert Dreyfus, talking in the BBC's Great Philosophers series in the early 1980s, points out:

> I remember in the film *2001: A Space Odyssey*, Stanley Kubrick has the robot HAL, when asked if he is happy on the mission, say, 'I'm using all my capacities to the maximum. What more could a rational entity want?' A brilliant expression of what anyone would say who is in touch with our understanding of being.
>
> We thus become part of a system which no one directs but which moves towards the

total mobilisation of all beings, even us, for their own welfare.

If HAL's comments are bleak, let's consider the 1969 dystopian hit *In The Year 2525* by Zager and Evans. The lyrics of the song seem to suggest the year 2525 might have come a bit early. In recent decades, telling the truth has become a revolutionary act. Once we obey the diktats of political correctness and toe the party line, it seems all will be okay in what we think, do, and say.

The lyrics continue: *In the year 4545/You ain't gonna need your teeth, won't need your eyes/You won't find a thing to chew/Nobody's gonna look at you.../Your arms hangin' limp at your sides/Your legs got nothin' to do/Some machine's doin' that for you.*

Heidegger wasn't optimistic about all this. He believed we might be stuck in the darkest night for the rest of human history. The only hope was that we would get involved in local concerns and that other meaningful events might make us appreciate non-efficient practices. We might even meet our future husbands and wives engaging in such activities, with no need to pick our offspring from the bottom of a long glass tube.

Professor Dreyfus again: "I think he has in mind such things as friendship, backpacking into the wilderness, running and so on. He mentions drinking the local wine with friends, and dwelling in the presence of works of art. All these practices are marginal precisely because they are not efficient." On theological eschatology, it is God and the Second Coming that will decide on the fate of humanity and the fate of the universe. All the transplanting of synthetic parts into humans won't deter extinction because of the heat death scenario. But before that happens, God will usher in a New Heavens and New Earth (1 Thess. 4.15-17; Rev. 21.1; 1 Cor. 15.51-52).

The crux of dystopian novels or songs points to the terror of existence if we become slaves-turned-cyborgs—if technology ultimately becomes our new god. In 2016, a radio talk-show host on The Kuhner Report told the story of how he and his wife and kids brought his pre-teen nephews on a cruise with them. Throughout the holiday, his nephews were constantly glued to their iPhones and refused to swim, play, talk, or engage in fun with their uncle, his wife, or children. He lamented with sadness how there would be no fun adventure memories of the trip for his nephews, who had continued to be addicted to their phones even on return home from the cruise.

This story is not uncommon. It's becoming the story of nearly every parent who sees a blue light flashing from their child's bedroom during the late-night/early-morning hours. All our technologies give us instant information, but the awe of the transcendent is blocked by these cold, technological devices.

The writer Patrick West said utopias will never happen. In an essay in *Spiked* magazine, he wrote:

> Only ideologues and the naive believe in fantasy lands. If people in Iceland, Denmark, Sweden and Finland are the happiest in the world, it's to do with the fact that they are, in that order, the highest consumers of anti-depressants in the world.
>
> Taking to drugs or to drink is a sign of wanting to escape reality, not embrace it. Humanity is imperfect . . . Only through accepting and embracing strife and woe as the

norms can we become better people and create a better world.

And, surely, it's better to accept and embrace strife than have our arms hanging limp by our sides while some machine is doing work for us? Such machines are expanding at a rapid pace in many industries worldwide.

Our countrysides are already hosting robotic 'workers' toiling away on farms, from dusk to dawn. These tend to the livestock's needs, but there's also plans to have them watch over crops, while other agri-robots hoe weeds and spray pests.

Experts claim that self-guiding machines will soon revolutionize farming and perhaps redraw some of our landscapes. In conclusion, this got me thinking of a story I once read in a comic magazine many years ago. I think it was in the Warren Publishing horror-fantasy magazine comics, many of which also featured Sci-Fi stories. To the best of my recollection (it was over 40 years ago), one such story had a peasant farmer living alone and exhausted from working on his dilapidated farm day and night.

He finds out that a plant, which grows in a nearby cave, can yield farm 'workers'. He goes to the cave and picks the plant then seeds it on his farm. The next day, he awakens to find hundreds of gremlin-like demons outside his house screaming for work.

Over a short period of time, he gets them to plough the fields, build extensions to his house, work the farm, and build a road to the village. But after they finish a task, they keep coming back to the farmer demanding more work.

In the end, the farmer sits alone in his now mansion, with nothing left for him to do. As he's slumped on his chair, the head demon approaches him and, once again, demands more work. The farmer says he's bored, adding: "Give me work." On that note, the rest of the demons storm the room and rip the farmer's limbs apart.

It seems the last human to leave the factory won't have to turn the lights off: leave them on for the robots to continue working—without ever taking a break. Finally, another great secular advantage about working robots—and the billionaires who control them—is they don't care about fairness, the minimum wage, or the Truth.

In the words of Huxley: "Great is truth, but still greater, from a practical point of view, is silence about truth." Truth is also linked to our past and the foundation of Western civilisation. But the foundations for Brave New Worlds are built on quicksand.

The Christian apologist Ravi Zacharias said that many years ago he was speaking at Ohio State University. "And I was taken to see the Wexner Center of the Arts," he said. "They wanted me to see it, and I wondered why. And when I walked into that building, I said, 'What is this building all about?' There are staircases that go nowhere. There are pillars that serve no purposes. And the man [showing me around] said, 'This is America's first postmodern building', and the architect said, 'If life has no purpose, why should our buildings have any design or any purpose?' So, he built it at random, without any purpose, as it were. I said, 'I have one question for you. Did he do that with the foundation as well?'."

At the end of the Sermon on the Mount, Jesus says: 'Whoever hears these words of mine, and does them, shall be likened to a wise man who built his house upon a rock: the rain descended, and the floods came, and the winds blew, and beat upon that house, and it fell not, for it was founded upon a rock.' [Matthew 7: 24-25]

Unfortunately, most of us have built our houses on quagmires. We want God to be what we want Him to be and not what He is. This is tragic, especially as the ends of our lives draw nearer.

As for a Brave New World drawing nearer: Kennedy, Lewis, and Huxley are now lying somewhere between Heaven and Hell. Those alive in the land of the living might live to see the day of Huxley's Godless vision. But only God knows the future—and He's not impressed with a brave new ship of fools. And where does it leave the devil as he watches the jaws of dystopian technology rise from the eternal seas to devour the crew? He's gonna need a bigger Hell.

Chapter 9

Is Solipsism the Weirdest Idea Ever?

Many years ago, I was doing research on the philosopher George Berkeley. I interviewed a Berkeley scholar who said two of his current students were solipsists (a person who thinks he or she is the only mind that exists). I asked him how he felt being viewed as the figment of a student's imagination, but he had no answer. I could have also asked him, 'Is it moral to murder a figment of your imagination?' but I didn't want to 'put a paranoid stone in his shoe' or have him look over his shoulder any time one of the solipsists entered the classroom.

In reality, there are more than likely no solipsists. The two students that the professor mentioned probably suffer from psychological nihilism. The common definition for this type of psycho-spiritual disease is extreme pessimism and scepticism; a moral relativism accompanied by being repulsed by existence while having a destructive nature with no loyalties. This corrosive state of mind is often associated with the German philosopher Friedrich Nietzsche, whose writing style and prose were outstanding, despite being Godless, deeply flawed with no system and doomed to failure.

Metaphysical-type Solipsism, as mentioned above, is a philosophical idea that a person thinks that he or she is the only existent thing. In other words, the 'external world' and all its contents are figments of 'your' imagination. By now you're probably thinking that solipsism is the weirdest concept in the history of philosophy and theology. But here's the problem: Try refuting it. There seems to be no way of either proving or disproving that solipsism is true or false unless one takes a leap of *reasonable* faith (not blind) and believes in God or that the external world is real and contains other minds.

For George Berkeley, who wasn't a solipsist but a Christian, everything is in the infinite Mind of God, the creator of us finite beings. In *Of The Principles Of Human Understanding*, he wrote: 'It is indeed an opinion strangely prevailing among men that houses, mountains, rivers and in a word, all sensible objects, have an existence, natural or real, distinct from their being perceived by the understanding . . . For what are the aforementioned objects but the things we perceive by sense? And what do we perceive besides our own ideas or sensations? And is it plainly repugnant that anyone of these, or any combination of them, should exist unperceived?'

The difference with someone claiming to be a solipsist, unlike God, is that 'he/she' has limited knowledge and power in the inner universe that he/she experiences. Then there's the question of death. What happens when the solipsist dies? One theory could be he/she is eternal, living life over and over again in an internal, cosmic, repeated loop.

Such a concept of cyclical patterns, but lacking eternity, has its origins in some Eastern philosophies; also, in *The Gay Science*, Nietzsche wrote about something similar but with eternal recurrence of all events that would mark the ultimate affirmation of life:

> What, if some day or night a demon were to steal after you into your loneliest loneliness and say to you: 'This life as you now live it and have lived it, you will have to live once more and innumerable times more; and there will

> be nothing new in it, but every pain and every joy and every thought and sigh and everything unutterably small or great in your life will have to return to you, all in the same succession and sequence' . . . Would you not throw yourself down and gnash your teeth and curse the demon who spoke thus? Or have you once experienced a tremendous moment when you would have answered him: 'You are a god and never have I heard anything more divine.'

Another spooky aspect of something resembling solipsism is lucid dreaming, which has often happened to me. This strange phenomenon occurs when one 'wakes up' *in* a dream. This is followed by a sense of panic, as the fear of not waking up and being trapped in the dream for days, if not years (a dream-time minute could feel like an awakened-time year). And the question must be asked: If everything in such a dream state seems real, then how does one prove that our ordinary awakened states are also real?

The Taoist parable, attributed to Chinese philosopher Zhuangzi (Chuang-tzu) (369 BC – 286 BC), of the butterfly dream, touches on reality vs. illusion:

> Once upon a time, I, Zhuangzi, dreamt I was a butterfly, fluttering hither and thither, to all intents and purposes a butterfly. I was conscious only of my happiness as a butterfly, unaware that I was Zhuangzi. Soon I awakened, and there I was, veritably myself again. Now I do not know whether I was then a man dreaming I was a butterfly, or whether I am now a butterfly, dreaming I am a man. Between a man and a butterfly there is necessarily a distinction. The transition is called the transformation of material things.

This short story ponders on how we know when we're dreaming, and when we're awake. What is real or illusion? Another anecdote was told by the great Christian philosopher Alvin Plantinga, when he gave a lecture in 2011 at Taylor University, Indiana. According to Dr Jay L. Wile, Plantinga started his talk with a few witticisms.

> . . . He talked about the philosophical view called Solipsism, in which one can only be sure that one's own mind exists. In the most extreme view of Solipsism, the Solipsist is the only real person that exists, and everyone else is simply a construct of the Solipsist's mind. He said while this might seem silly to many people, it is a serious issue in philosophy. He said that he once met a philosopher who was an extreme Solipsist. He was an elderly man, and everyone in his department took incredibly good care of him. They carried his books and materials when he had to go somewhere, made sure that he ate well, and generally watched out for his well-being. As Dr. Plantinga was leaving the Solipsist's department, he told someone else in the department how impressed he was that they took such great care of this elderly philosopher. The person replied, 'We have to take good care of him, because when he goes, we all go!'

Then there was the woman who allegedly said to the philosopher Bertrand Russell, who at the time was a solipsist (paraphrase), 'Mr Russell, I too am a solipsist, but I wish there was more of us.'

Funny anecdotes aside, the concept of solipsism is quite strange. Think about it: How do we know to be true with 100% certainty that the 'other' person has thoughts in his or her head? It seems Truth is the most important question in philosophy. With that in mind, I'll leave the last words to Jesus when He stood before Pilate: Therefore, Pilate said to Him, 'So You are a king?' Jesus answered, 'You say correctly that I am a king. For this I have been born, and for this I have come into the world, to testify to the truth. Everyone who is of the truth hears My voice.' Pilate said to Him, 'What is truth?' then he quickly left the room (John 18: 37-38).

Chapter 10

Do Spiders Dream of Arachnology Professors?

Have you ever wondered if insomniac spiders count imaginary arachnology professors jumping over college desks? Silly questions aside, it's not something that constantly occupies my mind. Anyhow, the other night as I rested in bed reading a book, I noticed a spider on the ceiling above where I lay. I wondered was it awake or asleep.

This sight occurs every seven or eight weeks, and if my ceiling was made from steel and I had a flame-throwing drone, I'd incinerate the creepy-crawly immediately because they give me the creeps. But instead, I usually get out of bed and, without killing the tiny creature, gently place the head of a sweeping brush under it in order to usher it out safely onto the window ledge.

Insects also give me the creeps even though they're different from spiders (spiders belong to the *class arachnida*, insects to the *class insecta*). And, on a micro level, the thoughts of one scaling down its silk web 'rope' like a Navy Seal during the middle of the night, and landing in my mouth as I sleep reminds me of what happened to John Hurt in the movie *Alien*, when he peeked into a pod.

However, what struck me again the other night when I reached the brush to the ceiling to catch the spider was the way it tried to dodge my efforts of capturing it by fleeing in the opposite direction. This, it seemed to me, is similar to how a human would react if a giant spider tried to do likewise to him or her. Being aware of anthropomorphising the situation, it nonetheless made me wonder: Do spiders, or insects, have some form of primitive consciousness? Or are such manoeuvring tactics nothing more than an electrochemical reaction in its brain?

In an article in internet's best scientific website, *Uncommon Decent*, Ronald R. Hoy, Cornell University professor of neurobiology and behaviour, considers the spider "one of the smartest of all invertebrates." But while its behaviour is comparable to that of many vertebrates, its anatomy is not, according to the article. It added: Dr Hoy and his colleagues wanted to study jumping spiders because they are very different from most of their kind. They do not wait in a sticky web for lunch to fall into a trap. They search out prey, stalk it, and pounce. "They've essentially become cats," said Dr Hoy.

There is certainly no doubting the spider's hunting prowess, especially their ingenuity in weaving complex web traps.

Such webs come in range of sizes, as do spiders. The smallest creatures, *Patu digua* from Colombia, are less than 0.015in in body length, while the biggest spiders are called tarantulas, with a body length of 3.5 inches and leg span of 9.8 inches. And if you're eating lunch, look away now: Cooked tarantulas are served as a delicacy in Cambodia!

But the ruthlessness of spiders devouring their prey in a web can make a person wonder what kind of a God would create such vile creatures, some of whom devour their tiny insect prey by injecting them with poison before eating them alive.

In an excellent essay last August in *Takimag* (*A Tangled Web*), Theodore Dalrymple posed the above theological question, after he witnessed such a killing by a funnel-web spider (*Agelina labyrinthica* to be precise) in his back garden.

He wrote that, in evolutionary terms, "we know that the funnel-web spider was fit to survive because it *did* survive, and we know that it survived because it was fit to survive. This seems to me to be dangerously circular, a pseudo-explanation. But what is the alternative explanation? That some force, divine or otherwise, created the funnel spider from a blueprint? This seems to me absurd. Who would decide that what the universe really needed for its completion was a funnel spider, let alone any or all of the other millions of arthropod species on earth?"

This is a profoundly complex question which relates to the study of theodicy and the Fall. There are various views by Christians as to how Adam's sin affected the animal kingdom. Were God's creations perfect before the Fall in the Garden of Eden? Did they develop sharp teeth and claws after the Fall or did they already have them? Some theologians believe that outside the bounds of the pristine Garden, there was a fallen world with death and destruction. Sin had entered the universe through the fall of Satan, and the earth was subject to his dominion. (I bet 2020 was one of his favourite years.)

If the Devil had already sinned before Adam and Eve, then evil had already been brought into the universe. Adam's sin, therefore, would affect only the earth, not the heavens; thus, some Bible commentators want to restrict Adam's sin to Eden – the 'Garden' that had not been tainted by sin. In such a tiny paradise, animals would have had to eat plants before the Fall.

St. Paul could not have been referring to sin affecting the *entire* world but instead only humanity (Romans 5:12). The Scriptures indicate animals were affected by the Fall. However, the new heavens and new earth will be free from all sin which testifies to the original state of creation without sin and death.

Furthermore, non-human creatures are not moral agents, thus they lack free will. And let's not forget, spiders that kill flies protect humans from the diseases that winged creatures bring to our kitchen surfaces. They also kill mosquitoes and other pests.

As for the suffering of the bigger creatures: In a fallen world, the animal kingdom is a place of perpetual screaming in the survival of the fittest. Have you ever seen a documentary on TV of a pack of hyenas eating a wildebeest while the poor creature is still alive, while the poor creature looks back at all his lower entrails hanging out of his body and being devoured? Such grotesque scenes make a spider's dietary habits look like a vicar's tea party in comparison.

However, to return to what goes on in spiders' brains: Besides my bedroom ceiling, there is another spider who has a nest in a tiny crack beside my car's rear-view mirror. It's bad enough having an ancient 17-year-old Saab without having a cobweb on the window. Anyhow, his (or her?) web is fixed to the mirror, which no doubt attracts insects or flies to land on the web. And even on the windiest day, if I'm driving at speed down the motorway, the web is like a net of steel in its durability; it never breaks.

But here's the difficult question: How did the spider know that the mirror on my car would attract its prey during daytime hours? How does it know how to weave the perfect 'net' to catch its lunch? And what is one to make of the effects drug-taking has on the design of a spider's web?

In 1948, Swiss pharmacologist Peter N. Witt experimented with such effects of drugs on spiders. He tested spiders with a range of psychoactive drugs, including LSD, amphetamine, mescaline, strychnine, and caffeine, and found that the drugs affect the size and shape of the web rather than the time when it is built. The drugs were administered by dissolving them in sugar water, and a drop of solution was touched to the spider's mouth. In some later studies, spiders were fed with drugged flies. The webs were photographed and this is what they looked like.

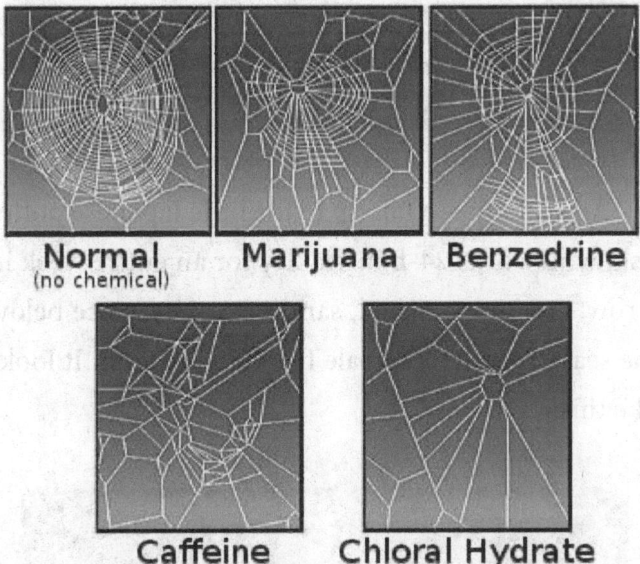

But webs aside, consider the other strange, sometimes ingenious, tiny creatures/organisms, like a colony of termites, that produce astonishing land structures. Most termites are blind and communicate by chemical, mechanical, and pheromonal cues; they build giant, complex mounds with sophisticated 'air-conditioning', as seen close-up in one of these two photos below:

And let's give a hat tip to the little Japanese puffer fish, who works 24-hours a day for an entire week in a row, creating a visual, sand-art masterpiece below the sea, to attract a female fish to mate with. It looks like this:

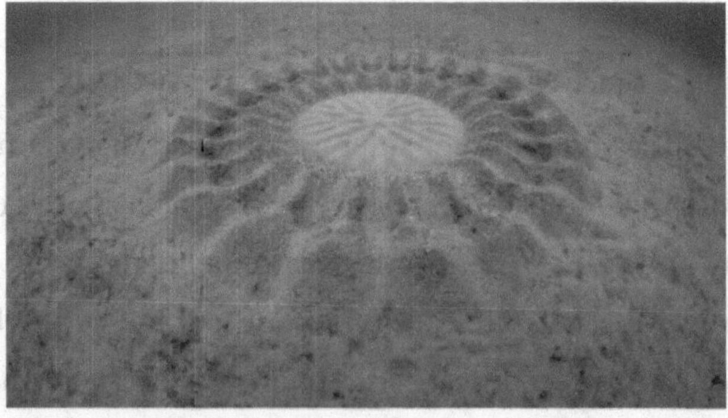

There must be some form of consciousness or invisible force of guidance at work here. How else could a fish be aware that the aesthetic beauty of a sand structure could attract a female to mate with him? Information of any sort has to come from a mind, an Informer.

In his new book *Miracle of the Cell* (2020), biochemist Michael Denton talks about the ways our bodies' individual cells appear, to researchers, to show intelligence: "No one who has observed a leucocyte (a white blood cell) purposefully—one might even say single-mindedly—chasing after a bacterium in a blood smear would disagree."

Although this sounds quite spooky, just like other non-human creatures or cells, it's unlikely such brainless organisms are conscious like us but instead are capable of reacting, as well as possessing, like a complex machine, information, like being hard-wired. But notice how all these things have one thing in common: They're purpose-driven.

So, my answer to 'Do spiders have consciousness?': In 1974, in a paper entitled, 'What is it Like to be a Bat?', the philosopher Thomas Nagel posed a similar question. The paper is a critique of reductionist theories of the mind. Nagel argues that consciousness has a subjective aspect, and that understanding other mental states is difficult or impossible for those not able to experience those mental states.

Nagel chose bats because they are mammals, and relatively closely related to humans. But he said that "anyone who has spent some time in an enclosed space with an excited bat knows what it is to encounter a fundamentally alien form of life." Just like spiders, I'd imagine, Nagel writes that bats experience a range of activity and possess a sensory apparatus so different from ours that imagining what their experiences are like is an extremely complex question.

But let me briefly distinguish different states of consciousness: Like evolution, which seems plausible on the micro level because nearly every creature evolves ever-so-slightly and slowly in some way or other, but it is not the same as Darwinian Evolution, which claims extraordinary leaps of change, e.g. from a sponge to Michelangelo's painting *The Creation of Adam* on the Sistine Chapel ceiling. Consciousness can also be simple and basic to something more profound like awareness of the self: Qualia (the awareness of the "I"). Can you see where this is going? You got it—next level, the Soul. And it's hard to imagine spiders have qualia or souls.

Even without recourse to the notion of a soul, there are several ways in which the mind is not simply brain. The wet, grey organ known as the brain does not have mental states such as love, hate, or sadness even though they interact with the brain as they are expressed in the natural world. Then there is the problem of propositional attitudes such as fear, hope, desire, wish, dread, and regret. The last thing my ceiling spider was thinking was: "Oh no! ***I*** hope he doesn't kill ***me. I*** better run for cover because ***I'm*** too young to die!"

As to where the mind resides, that is the biggest mystery in philosophy. Although it interacts with the brain, it can't be a kind of invisible vapour hovering above one's head. Nor can it be located in some part of the universe, as it's supernatural and outside of space and the material world, even though its thoughts manifest itself in the world, as mentioned above.

If epiphenomenalism is true (mind is brain), then how come fake drugs can work in placebo effects?

Also, how can parents' love of their children be nothing more than an electrochemical reaction in the brain? And at what point in evolution did the atoms in brains develop morals? At what point did a wet slab of blubber, lodged in our skulls, belch out consciousness? That we can have logic, reason, and truth evolving out of a material process that is aimless, purposeless, misguided, and unaware of self seems absurd. My resident creepy-crawly is not likely to weave a web on the ceiling with the words, "Please don't kill me".

Chapter 11

Agnosticism, Appearance, and Reality

When we hear or read about the word 'agnosticism,' many people tend to believe that it refers exclusively to scepticism on belief in the existence of God. But that's not the case. People can be agnostic on many things, such as historical events, like the authenticity of the moon landings, who shot JFK, or even the existence of the external world.

In my youth, I was an agnostic sitting on the proverbial fence on all things theological, but I eventually jumped off when my cosmic rear end began to cause me much pain (the origin of that pain came from my disordered soul, which I didn't believe existed back then).

So, I want to concentrate, in a general way, on an example of how agnosticism can be seen as a kind of knowledge claim and not a cautious, rational scepticism. I also want to add a geo-spatial component to the essay linked with agnosticism as to how we perceive the external world in both its appearance and reality and how that shapes our beliefs.

The American writer, public speaker, and documentary maker Dinesh D'Souza is a man who has his worldview solidly based in reality. He also comes across as a highly intellectual, decent man. Over the years, I watched most of his speeches and debates and found him to be dynamic in his delivery, fair and balanced, and quite witty with a funny sense of humour.

Lately, on a regular online slot, he has a podcast, where he muses on social and current affairs both domestically and internationally. His views are mostly informative, and I enjoy watching his videos.

However, in his recent January 25 podcast, he spoke of agnosticism and how many believers in God are in some way agnostic. He said the following, while also commenting on faith, belief, and knowledge (edited version):

> "... *You have to make a distinction in believing something and knowing something. A belief is something that you hold but it stops short of knowledge which is something that you know definitively and for sure without a shadow of a*

doubt... It doesn't take any faith for me to believe in trees; it doesn't take any faith for me to believe in my wife standing here in front of me and producing this podcast. I don't have to believe in Debbie [his wife], because I know her; she's right here. Belief is one step short of knowledge and it creates room for faith..."

Dinesh says "it doesn't take faith for me to believe in trees" or belief in his wife standing in front of him. Really? Can one have certitude of the reality of the external world? Could that 'tree' Dinesh sees or his wife, Debbie, be figments of his imagination? I don't believe they are but, according to solipsism, which is extremely hard to refute, both 'tree' and 'Debbie' might not exist externally but only in the mind of Dinesh. In philosophy, this is called the problem of other minds. (For the record, I find solipsism both fascinating and disturbing, but to repeat, I believe Debbie, Dinesh, and trees exist, and that's based on reasonable faith, not blind faith.)

There seems to be no way of either proving or disproving that solipsism is true or false, unless one takes a leap of *reasonable* faith (not blind) and believes in God or that the external world is real and contains other minds. It seems the only certitude is, "I think, therefore I am," to quote Descartes. That is, you can be certain that you definitely exist.

Dinesh was responding to comments made by astrophysicist Neil deGrasse Tyson, who says he claims the title "scientist" above all other "…ists," despite his being "constantly claimed by atheists" but says he is not and that he's an agnostic. But if de Grasse Tyson says he's agnostic, why does he, like most agnostics, believe, without a shred of scepticism, anything that scientists say about science?

Professor John Staddon writes in *Science in the Age of Unreason* that "many social scientists have difficulty separating *facts* from *faith*, reality from the way they would like things to be. Critical research topics have become taboo which, in turn, means that policy makers are making decisions based more on ideologically driven political pressure than scientific fact" (xii).

Also, in reports regarding NASA's cosmic achievements, agnostics shed little doubt on the authenticity of such claims. Think about it: All the news we get about space exploration, moon, or Mars landings comes from the secondary source of the mainstream media. And if the year 2020 and the internet told us anything, a lot of media reportage is leftist propaganda, false and misleading.

But don't take my word for it: John Swinton was the former Chief of Staff for the *New York Times*. In 1953, he was asked to give a toast before the New York Press Club. He said that there is no such thing, at this date of the world's history, in America, as an independent press. Swinton added (edited version):

> *"You know it and I know it. There is not one of you who dares to write your honest opinions, and if you did, you know beforehand that it would never appear in print. I am paid weekly for keeping my honest opinion out of the paper I am connected with... We are the jumping jacks, they pull the strings and we dance. Our talents, our possibilities, and our lives are all the property of other men. We are intellectual prostitutes."*

But I slightly digress. At least in the case of God, the evidence for fact and truth on creation and morality is everywhere: DNA; Fine-tuning of the universe; *Objective* moral values and duties; Intelligent design, etc. So, to return to agnosticism: Is it really neutral? And can one be agnostic about their agnosticism (*ad absurdum*)?

Firstly: It's debatable that agnosticism is a neutral stance on whether or not God exists. I tend to see it as a truth claim in that the agnostic admits to know his or her agnosticism is the right worldview regarding theism; otherwise, they wouldn't be an agnostic. And what's right for them must also be true for them, and what's true must be a knowledge claim, despite being dressed up in 'healthy scepticism'. (I'm not referring here to being agnostic on trivial issues.)

For the philosopher George Berkeley, who wasn't a solipsist but a Christian, everything is in the infinite Mind of God, the creator of us finite beings. In *Of The Principles Of Human Understanding,* he wrote: "It is indeed an opinion strangely prevailing among men that houses, mountains, rivers and in a word, all sensible objects, have an existence, natural or real, distinct from their being perceived by the understanding . . . For what are the aforementioned objects but the things we perceive by sense? And what do we perceive besides our own ideas or sensations? And is it plainly repugnant that anyone of these, or any combination of them, should exist unperceived?"

As for agnosticism: A similar worldview is scepticism. But here's the problem: Is the sceptic sceptical of his scepticism? And if so, is he sceptical of his scepticism about his scepticism? (Ad absurdum).

Then there is the problem for the agnostic of appearance and reality. Those trees that Dinesh mentions can be visually perceived in more ways than one, as can his wife, Debbie, which throws out the question: Are agnostics agnostic about appearances in the external world? They should be if they want to be consistent.

However, a tree seen from many angles looks different, as does a human being. This geographical phenomenon by visual degrees can also be seen in everyday domestic objects, the countryside landscape, or works of art. From a distance of thousands of miles, Jupiter's ice-covered moon Europa looks like a white snooker ball; but on closer inspection it vaguely resembles a Jackson Pollock abstract painting with its fractured crust and zig-zag patterns.

But also, as an example, consider the great painting of the *Girl With a Pearl Earring* by Dutch artist Johannes Vermeer. This most-recognised artwork was recently subjected to two years of detailed research which has revealed strange-looking details visible under a microscope.

The surface condition of the painting (which has been subject to multiple restorations over the years) was studied by researchers Emilien Leonhardt and Vincent Sabatier, of Hirox Europe using a custom-mounted microscope. The resulting 10-billion-pixel panoramic scan—a gigantic stitch of 9,100 individual microscopic photographs, is seen below in 2D and 3D. What's fascinating is the light in the girl's eye and its surrounds.

The British philosopher Bertrand Russell once said that to view such images from a convenient/common distance is a kind of favouritism on the person viewing it. I assume that distance is usually two-feet to 30ft (sometimes miles), depending on the object, be it a person, table, building, aeroplane or mountain. In the painting below, it's usually two to four feet; some 50ft to 100ft away might make the image look like a stain on a wall. But under a microscope the appearance begins to break down into something that's unrecognisable (see images below of Vermeer's painting, as mentioned above).

Painting viewed at a 'favourable' distance (notice the light in her eyes).

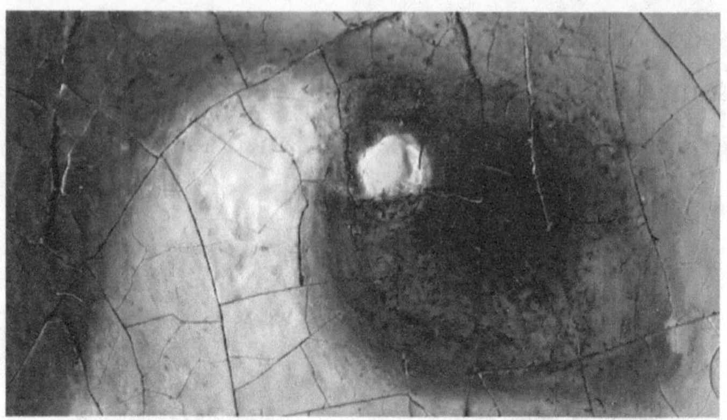

Close-up of girl's eye at a less-favourable distance to the human visible perception.

Macabre-looking micro close-up of light in the girl's eye (between the limbus and sclera), resembling a discarded clump of oyster meat or a poached egg on a cracked desert floor.

Also, consider these other images of snowflakes.

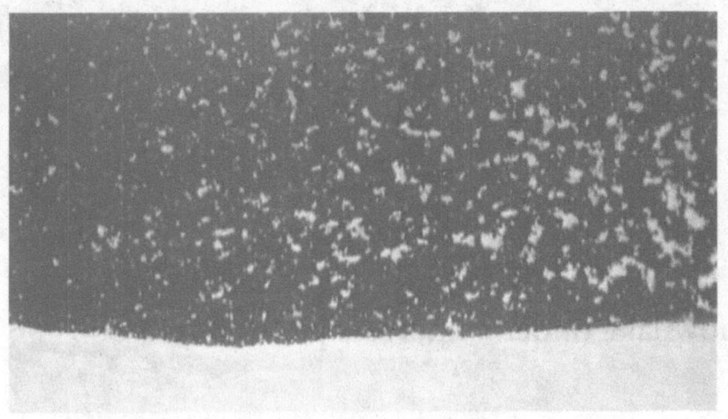

Above, snow falling outside a windowpane.

A snowflake close-up.

Snowflake under a microscope.

Snowflake under an electron microscope. Which one looks like the 'real' snowflake?

Then there is the phenomenon of optical illusion. This is where a person's visual system differs from reality in certain visual categories, where sometimes light plays a powerful role. One example of an optical illusion is that of boat's oar apparently 'bending' while half immerged in water. Below are two other examples:

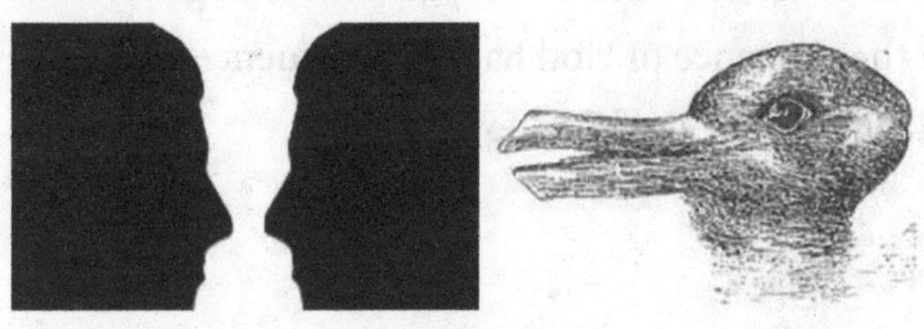

Reversible figures and vase (or the figure-ground illusion) and the Rabbit-Duck illusion.

In an essay in *The Imaginative Conservative* (Dec 9, 2017), Thaddeus Kozinski said we must keep our eyes fixed on the suffering and death of Jesus Christ and hold on to him for dear life in the midst of the evils that would otherwise suffocate our souls and eclipse all the beauty and love that this world still contains. He added: "Agnosticism is the ultimate stupidity and wickedness because it doesn't so much reject God as ignore him. If I were God, I'd be more angry at such cold indifference than anything else."

I'd go a little further than Kozinski on the dangers of agnosticism regarding scepticism on belief in God. Being agnostic about the existence of Big Foot, the Loch Ness Monster or life on Mars, etc., will not have any profound consequences either in this life or on your eternal soul when you die. But agnosticism on the existence of God has a consequence that's unimaginably bleak if God exists.

Chapter 12

Ever Since Derrida

We live in a world where words are incrementally being corrupted, obfuscation widely practised, double-talk and blatant lies emanating every day from the mainstream media, scientists, academics, clerics, and politicians. Some evil acts are now called good; dark is called light; and bitter called sweet. But there's nothing sweet about the Age of Confusion that is the 2020s. One wonders how we got to this bleak situation of syntactical, semantical anarchy.

About 25 years ago, while lecturing in a university, I attended a talk one evening by Jacques Derrida, who lived in France but was born to Jewish parents in Algiers in 1930 (he died in 2004). The talk was scheduled in the opposite building in the college campus.

A philosophy professor I knew was a close friend of Derrida, and he invited him to give the lecture. Being deeply interested in why Derrida had rock-star status and was regarded as the world's most-famous philosopher, I decided to see what he had to say. On the night of the talk, I managed to sit through the entire convoluted waffle in great distress, while the audience clapped like circus seals when the talk ended.

I wanted to walk up to Derrida, who came across as a stereotypical French philosopher with a perfect tan, and say, "Mr Derrida, can you not communicate deconstructionism in a more understandable, clearer way? I work in the opposite building teaching prose, syntax and semantics, and my motto to my students is taken from the Bible: "But let your 'Yes' be 'Yes,' and your 'No,' be your 'No.' For whatever is more than these is from the evil one." (Matthew 5:37).

I could've also quoted him Isaiah 5:20, but Derrida would probably have challenged the merits/qualities of the different shades (or hierarchies) of dark and light, likewise the endless, or multiple, variations of the degrees of bitter and sweet. But with numerous possibilities, to what end, i.e., ad absurdum?

For Derrida, language is impossible to determine. Ironically, he probably was aware of his self-defeating words: if language is impossible to determine, then such a statement is self-refuting. I doubt if the audience on the night understood the lost-in-translation French-to-English versions of their hero's work. I mean, who could fail to not understand statements like, "To see to it that the beyond does not return to the short-of is to recognize in the contortion the necessity of a trail"? (*The Madness of Jacques Derrida*, by Josh Herring). Derrida wasn't mad, but I found him to be a poor verbal communicator of deconstructionism, which is an interesting concept similar to the philosophical ideas of appearance and reality.

If I can borrow from the English Franciscan friar, William of Ockham (1287-1347), and his principle 'Ockham's Razor', I believe that the method of deconstruction should not be multiplied beyond necessity. In other words, one doesn't need to analyse language or sentences into the ground to understand their true, absolute meaning. When presented with competing meanings in language, one should select the concept with the fewest assumptions and makes the best common sense; otherwise, we're faced with an analytical, textual probe into endless possibilities.

In fairness to Derrida, he's not the worst offender of poor communication. Pope Francis would give him a run for his money with his infamous airplane gobbledygook press statements. Who to blame? Wittgenstein? Heidegger? Hegel? The art of talking from both sides of one's mouth and mangling complex concepts has a long history. And there's also a fear in questioning the emperor's new clothes.

Some 25 years ago, consider the *Sokal Affair*, also called the *Sokal Hoax*. It was a scholarly hoax performed by Alan Sokal, a physics professor at New York University and University College London. In 1996, he submitted an article to *Social Text*, a 'revered' journal of postmodern/cultural studies. The submission was an experiment to test the journal's standards of rigorously analysing text for original, authentic material. However, the paper was a spoof, riddled with nonsense. It was called *Transgressing the Boundaries: Towards a Transformative Hermeneutics of Quantum Gravity* and was published in the journal's spring/summer 1996 "Science Wars" issue. It was alleged at the time that leftist, anti-intellectual sentiment in liberal arts departments (especially English departments) caused the increase of deconstructionist thought, which eventually resulted in a deconstructionist critique of science. So much for deconstructionism.

So, to repeat, how did we get to this bleak situation? Dwight Longenecker recently wrote in *The Stream* (Nov 17): "It's not too hard to figure out what has gone wrong. It's called relativism. Relativism is the idea that there is no such thing as truth, or if there is, it is impossible to state the truth accurately and authoritatively. In other words, dogma—the definite expression of truth—is impossible. Relativism has been creeping into our society like an insidious cancer for the last 70 years."

Deconstructionism reminds me of relativism and the parable of the blind men and an elephant. This is a story of a group of blind men who have never come across an elephant before and who learn and imagine what the elephant is like by touching it. *Each blind man feels a different part of the elephant's body, but only one part, such as the side or the tusk. They then describe the elephant based on their limited experience and their descriptions of the elephant are different from each other. The moral of the parable is that humans have a tendency to claim absolute truth based on their limited, subjective experience as they ignore other people's limited, subjective experiences which may be equally true.*

In other words, the 'whole' is not the 'holes' (e.g. tiny porous gaps, like inside a tree, etc, making up the 'whole'). Despite this, common sense and logic should lead the way in interpreting language and the world around us. Failing to believe in an objective reality is to be trapped in a solipsistic swamp of secular solitude.

Chapter 13

The Confusion of Buridan's Ass

I'm going to look at the philosophical paradox of *Buridan's Ass* and where some confusions in its conclusion might lie. But first I want to briefly comment on an essay I wrote recently on why I believe the brain is not the mind, which posed the following question: At what point in evolution did the atoms in brains develop morals? A thoughtful reader of the essay replied with the following comment:

> *When man discovered he could choose. In situations where two choices are not equal, when one is ostensibly better, that is more life-affirming than the other, and being cognizant that one choice is better than the other, choosing, that is doing what is right, became synonymous with the moral imperative.*

My reply to the reader was:

> *On Naturalism, humans (and other creatures) never 'discovered' they could choose, as they were always endowed with instinctual decision-making attributes based on appetite or survival requirements. Similarly, robots can 'decide' when 'choosing' to make a move playing chess or other functions. The input of the information into the system [by humans], will influence the moves it makes. But my question is on the existence of freedom of the will and choosing morally, based on objective moral values and duties. On atheism, atoms just blindly bump into one another. Also, be careful using terms such as 'ostensibly better,' 'one choice better than the other,' or 'doing what is right' (on atheism, says who?). Such terms, subjectively, bring great comfort to the psychopaths of this world. (*'The Brain is not the Mind,' *New English Review,* March 2022*)*

This brings me back to the paradox of *Buridan's Ass*. I sometimes wonder is the confusion to this paradox due to a category error, like the comment above on the mind/brain problem, as well as confusion in the term 'identical'. For those who have never heard about it: *Buridan's Ass* is commonly considered to be an illustration of the problem of acting rationally, 'choosing', or being endowed with freedom of the will.

Named after the 14th-century French philosopher Jean Buridan, but probably originally formulated by ancient Greek philosophers (possibly Aristotle), it goes something like this: Imagine a hypothetical situation where a donkey is both equally hungry and thirsty and, in terms of location, he is placed *equidistant apart, midway between water and food. At this point the paradox assumes that the ass will always go for that which is closer; or does it starve to death due to hunger and thirst because of its paralysis in not being able to 'choose'? Put simply: It allegedly cannot make any rational decision between the water or food due to the seemingly exact distance of both? (*The definition of 'equidistant' is that of equal distance [in most ways], and not 'identical.')

But distance aside for a moment, is this a cognitive illustration of a creature lacking high-consciousness reasoning or free will? When philosophers or theologians talk about rationality or free will, they usually refer to it as a moral question, not a preference/decision-making one; thus, it seems that *Buridan's Ass* might be confusing preferences or decision-making an ass might make ('choosing'), as opposed to a moral (or deeply rational) decision that he can't make because animals are not moral agents but instead are meat machines made up of billions of atoms in a state of flux. Is this the same for all species?

Let's briefly look lower down the scale regarding insects: Researchers are beginning to understand how, without much in the way of individual decision-making power, ants can make complex decisions, according to a recent article in *Mind Matters* (March 30). It's best understood, they say, as something like an optimization algorithm: Scientists found that ants and other natural systems use optimization algorithms similar to those used by engineered systems, including the Internet.

In his book *Miracle of the Cell* (2020), biochemist Michael Denton talks about the ways our bodies' individual cells appear, to researchers, to show intelligence: "No one who has observed a leucocyte (a white blood cell) purposefully—one might even say single-mindedly—chasing after a bacterium in a blood smear would disagree." Did it decide to chase?

Although this sounds quite eerie, just like other non-human creatures or cells, it's unlikely such brainless organisms are conscious like humans but instead are capable of reacting, as well as possessing, like a complex machine, information, like being hard-wired. But notice how all of these things have one thing in common: They're purpose-driven.

But back to animals: Although I believe they have lower levels of consciousness, despite having no sense of self, it's highly unlikely that they think rationally. Cameron Buckner, assistant professor of philosophy at the University of Houston, argues in an article published in *Philosophy and Phenomenological Research,* that a wide range of animal species exhibit so-called "executive control" when it comes to making decisions, consciously considering their goals and ways to satisfy those goals before acting.

Buckner acknowledges that language is required for some sophisticated forms of metacognition, or thinking about thinking. But bolstered by a review of previously published research, Buckner concludes that a wide variety of animals—elephants, chimpanzees, ravens and lions, among others—engage in rational decision-making. (Jeannie Kever, University Website, Nov 1, 2017)

As for the paradox of the hungry and thirsty ass: Buridan concludes that a rational choice cannot be made in the donkey's scenario and that, until it can be debunked or clarified, we must suspend judgment. Is he right? Or did he overlook the true definition of the word/concept 'Identical'? Can two separate things *be* identical? To degrees, in some ways they can, but not in every conceivable way.

In Christian theology, in most ways, God the Father is identical to the Son and Holy Spirit (difference is found in relation), but uniquely and ultimately, the Three Persons are One. Regarding humans: A finite person, and his or her identity, is twofold and not identical. His or her status is that of a) the species homo sapiens, and b) the separate individual him/herself, i.e., their human character(s): Both physical and spiritual beings.

So, if no two things are separate in every single way right down to the micro or geographical level, then *Buridan's Ass* does not have identical options of choosing the food or water. The very fact that the food and water spatially occupy two different left/right geographical locations, also reinforces such a notion of 'difference,' as opposed to 'identical.'

There are lots of other examples in everyday life of how things can't be identical: 'Identical' twins are not 100% the same. While environmental factors can, and usually do, alter their appearances, do their DNAs differ? According to assistant professor of genetics at the University of Pennsylvania, Ziyue Gao: "One particularly surprising observation is that in many twin pairs, some mutations are carried by nearly all cells in one twin but completely absent in the other." (*Now*, scientific journal article by Tracy Staedter, June 14, 2021.) Then, there are so-called 'identical' bank notes or coins, but try looking at them under a microscope. The key word in the confusion regarding the paradox of *Buridan's Ass*, should be 'almost.'

Chapter 14

Understanding Phenomenology

"Hi, my name is Steve, let's talk about phenomenology."

That would be a bad chat-up line for any single man to say to a woman, despite it being an interesting concept worth discussing. Interestingly, the person who utters such words would see the 'object' of his desire as more than a sack of molecules in motion, but instead someone endowed with beauty in motion as well as, hopefully, having a loving heart.

On first sight, he would see her immediately in a phenomenological way and not scientifically. I refer here, of course, to the primary world of phenomenology and not the clown subculture we have to occasionally endure on Planet Woke, where genders should not be assumed and chatting up a woman is retardedly perceived by deeply unhappy people as a psychological sexual assault.

In the real, sane world, phenomenology as a way of perceiving reality, was developed earlier in the

20th century by the philosopher Edmund Husserl. Besides men admiring attractive women in bars, the subject has many theological implications in biblical hermeneutics, as well as how the average person's consciousness perceives the reality of the external world.

Some 14 years ago, I interviewed the world's leading scholar in phenomenology, Professor Dermot Moran. What struck me about Professor Moran was his lucidity when describing philosophical matters, as well as his being quite a pleasant chap to converse with.

I had just read his big book, *Introduction to Phenomenology*, which was described by David Bell from the University of Sheffield as "the most accessible, the most scholarly, and philosophically the most interesting account of the phenomenological movement yet written." With this in mind, I wanted to incorporate some of the ideas into a book I was writing on the Irish philosopher, George Berkeley.

I had already interviewed a Berkeley scholar at Trinity College about Berkeley's Idealism but wanted to expand my knowledge of phenomenology. For the lay-person, phenomenology can be a little complex, but I hope to make it as accessible as possible in the following paragraphs by giving a few examples.

Let us start with the objective reality of the external world and how we perceive it. Scientifically (secondary), we describe it quantitatively, whereas phenomenologically (primary), we describe it in an immediate way without any deep analysis. An example of phenomenology is when we see the sun rise or set, while an example of the scientific method describes the earth tilting for both sunrise and sunset.

Another example of the scientific method is the description of a tree and the various components which it contains: molecules, atoms, its shape, height, weight, age, etc. On phenomenology, our immediate perception is that of a beautiful tree with a trunk, branches, and leaves. We might even want to sit under it for shelter from the rain or shade from the sun. The last thing we want to do is measure it, unless, of course, one is a tree surgeon or a botanist doing research.

In our everyday lives, we certainly don't use the scientific method for the majority of our actions. When 'putting the kettle on' for a cup tea, we don't think of boiling the H_2O till the molecules move around and exceed the strength of the hydrogen bonds between the molecules, causing them to separate from the other molecules in order to increase the thermal energy to a heat point of 100^{C} (212^{F}). I can't remember the last time I made tea, pondering over the intermolecular bonds breaking up as the heat increased the kinetic energy of the system, causing the molecules of the water vapor to move faster with the temperature increasing.

In contrast, phenomenologically, our primary truths and experiences, if functioning properly, are sufficient to tell us all we need to know about an action, event, or object. Afternoon tea aside, consider a night at the opera or a day at the races: At the opera, we hear the orchestra playing, while sopranos and tenors sing to rapturous applause, while at the races, we see horses gallop and people cheer, all the while experiencing our five senses of sight, hearing, touch, taste and smell.

However, scientifically, the aforementioned events are a different story of a secondary experience, similar to the boiling water mentioned above; despite this, we can still know reality by its phenomenological truths.

Why is all this important, you might ask yourself. I believe it's important to understand what it is that gives both us and the material world of objects its fundamental structures and mechanisms. Which brings me to theism.

A retired Christian professor of philosophy with whom I regularly correspond briefly mentioned that many Catholics don't have an 'adult faith.' The term for such a belief system is 'properly basic' or 'fideism.' Liking it to phenomenology, such belief is primarily existentially immediate, whereas a more analytical belief is based on hard evidence, similar to the scientific method, based on investigation and/or reasonable faith as opposed to blind faith.

The British philosopher John Locke (1632-1704) made the distinction between ideas, immediately aware in perception to people, and material things caused by such ideas. This distinction runs counter to George Berkeley's philosophy, in that ideas are the first (most immediate) things that spark an awareness in human beings, as in sound, touch, sight, smell, and so forth.

But Locke was trying to find out what is it that is 'out there', whereas Berkeley, who was more of a phenomenalist, looks from within as his starting point. Chairs, tables, trees and everything in the so-called material world are not active in the same way as minds are. Berkeley is confident in the knowledge that such objects, unlike minds, are not perceiving and do not have ideas (that's panpsychism). Everything, he rightly claims, is in the infinite Mind of God.

In *Of the Principles of Human Understanding*, Berkeley wrote: "It is indeed an opinion strangely prevailing among men that houses, mountains, rivers and in a word, all sensible objects, have an existence, natural or real, distinct from their being perceived by the understanding … For what are the aforementioned objects but the things we perceive by sense? And what do we perceive besides our own ideas or sensations? And is it plainly repugnant that anyone of these, or any combination of them, should exist unperceived."

The ancient Hebrew writers, inspired by the Holy Spirit, had to communicate the world and the cosmos in a language that could be immediately understood. To explain things the way Berkeley did or in the scientific method would greatly confuse the